T0177607

Microbiology of Infectious Disease

Microbiology of Infectious Disease

Integrating Genomics with Natural History

Sandy B. Primrose

Independent Biotechnology Consultant, High Wycombe, UK

OXFORD
UNIVERSITY PRESS

OXFORD
UNIVERSITY PRESS

Great Clarendon Street, Oxford, OX2 6DP,
United Kingdom

Oxford University Press is a department of the University of Oxford.
It furthers the University's objective of excellence in research, scholarship,
and education by publishing worldwide. Oxford is a registered trade mark of
Oxford University Press in the UK and in certain other countries

Published in the United States of America by Oxford University Press
198 Madison Avenue, New York, NY 10016, United States of America

British Library Cataloguing in Publication Data
Data available

Library of Congress Control Number: 2021949439

ISBN 978–0–19–286384–3 (hbk)
ISBN 978–0–19–286385–0 (pbk)

DOI: 10.1093/oso/9780192863843.001.0001

Printed and bound by CPI Group (UK) Ltd, Croydon, CR0 4YY

Cover image: poba/Getty Images [front]. Angelina Bambina/Shutterstock.com; ImageFlow
/ Shutterstock.com; Kateryna Kon/Shutterstock.com; solarseven/Shutterstock.com.
[back & spine].

Preface

When I was an undergraduate student in microbiology in the early 1960s the teaching was very descriptive and much of the time was spent on detailing the properties of bacteria. The natural history approach being pioneered by C.B. van Niel at the Pacific Grove Marine Station in California had not made its way to Scotland at this time. So, for me, it was a great joy one day to discover Kenneth Thimann's book *The Life of Bacteria* (2nd edition, Macmillan, 1963). As a post-graduate student at the University of California, Davis I was exposed to a much wider view of microorganisms, including bacterial diversity and bacterial physiology. Over the following years I noted a trend towards investigating how microorganisms behaved in their natural habitat as opposed to laboratory culture and at the University of Warwick where I was an academic, this very much was the focus of our teaching and research.

In the late 1990s, the first complete sequences of two bacterial genomes were published and created much excitement among microbiologists. From the analysis of sequence data, it became possible to elucidate all the biochemical reactions that the bacteria possibly could undertake and all the molecules that they possibly could synthesize, many of which had hitherto been unknown. Now, a bacterial genome can be sequenced in a matter of hours and with the help of a panoply of software programmes the inner workings of bacteria can be probed in great detail. As more and more genomes were sequenced it was realized that the classification of bacteria on the basis of shared properties was erroneous and so bacterial taxonomy underwent constant revision. This led to a major growth in DNA-based phylogeny, the study of relationships among different groups of organisms and their evolutionary development. Today, scientific papers on bacteria carry extensive and detailed analysis of DNA sequences, usually accompanied by phylogenetic trees, but it is difficult to gain insights into the biology of the organism as a whole. Microbiology has almost become a branch of molecular biology with the biology getting lost in the molecular detail.

My objective in writing this book is to cut through the molecular information overload and put the new sequence-derived information in the context of the natural history of the organisms. That is, to tell a story and get a sense of what the organism does and how it interacts with its environment. Classical microbiologists, if such people still exist, might be puzzled by the choice of organisms. However, the decision about which organisms to cover and not to cover was made on the basis of whether or not there was a different story to tell. The fact that there are so many different stories, but with underlying themes, is what makes microbiology so interesting.

The reader might be puzzled by my use of the full Latinized name of each organism at all times. This has been done because it makes reading the text much easier. Similarly, I have minimized the use of abbreviations. Also, molecular analysis has brought with it a whole new lexicon which confuses the non-specialist reader. I deliberately have chosen to ignore this jargon where possible.

Sandy B. Primrose
September 2021

Contents

Contents

Glossary

Bacteraemia A bacterial infection of the bloodstream

Endemic A disease that is constantly present in a population or particular area but only affecting a small number of individuals. An outbreak is a greater-than-anticipated increase in the number of endemic cases which if not quickly controlled can become an epidemic.

Enzootic A disease that is constantly present in an animal population but only affecting a small number of animals at any time.

Epizootic The appearance of a particular disease in a large number of animals in the same place at the same time.

Founder effect Loss of genetic variation and/or fixation of random mutations in a population. Founder effects occur when a new population is established by a small number of individuals (founder population) randomly derived from a larger ancestral population.

Horizontal gene transfer The movement of genetic material between unicellular and/or multicellular organisms other than by the ('vertical') transmission of DNA from parent to offspring (reproduction).

Housekeeping genes Genes that are required for the maintenance of basic cellular function and are expressed in all cells of an organism under normal and patho-physiological conditions.

Genome reduction The process by which an organism's genome shrinks relative to that of its ancestors.

Genomic island Part of a genome that has evidence of horizontal origins (cf). A genomic island associated with pathogenesis is known as a pathogenicity island.

Lysogenic conversion A change in the properties of a bacterial cell as a result of its infection with a temperate bacteriophage.

Nosocomial A nosocomial infection is one acquired in a hospital.

Pan-genome The entire set of genes for all strains of a particular species. It includes the core genome containing genes present in all strains, the accessory genome containing 'dispensable' genes present in a subset of the strains, and strain-specific genes.

Pathoadaptation   The genetic changes that occur in an organism to increase its pathogenicity.

Pathogenicity island   A genomic island (cf) containing genes associated with pathogenesis.

Pathovar   A bacterial strain or set of strains with the same or similar characteristics, that is differentiated from other strains of the same species or subspecies on the basis of distinctive pathogenicity to one or more hosts.

Phylogeny   The study of relationships among different groups of organisms and their evolutionary development.

Population bottleneck   Major reductions in the population sizes of organisms.

Pseudogene   Nonfunctional segments of DNA that resemble functional genes. They lack regulatory sequences needed for transcription or translation or have coding sequences that are obviously defective due to frameshifts or premature stop codons.

Spillover   The transmission of a pathogen from a vertebrate animal to a human.

Zoonosis   An infectious disease caused by a pathogen that has jumped from a non-human animal (usually a vertebrate) to a human.

List of Abbreviations

aDNA	Ancient DNA
AHL	Acyl homoserine lactone
AIDS	Acquired immunodeficiency syndrome
CAI	Community-acquired infection
CDAD	*Clostridium difficile*-associated disease
CDC	United States Centers for Disease Control
cgMLST	Core genome multilocus sequence type
CI	Cytoplasmic incompatibility
ECM	Erythema chronicum migrans
ELISA	Enzyme-linked immunosorbent assay
ETI	Effector-triggered immunity
ETS	Effector-triggered susceptibility
GC	Guanine + cytosine
GDP	Guanosine diphosphate
GM	Genetically modified (crops)
GTP	Guanosine triphosphate
HAI	Healthcare-acquired infection
HIV	Human immunodeficiency virus
HTS	High throuput sequencing
IS	Insertion sequence
Mb	Megabase (1 million bases)
MDR	Multi-drug resistant
MERS	Middle East respiratory syndrome
MGE	Mobile genetic element
MLST	Multilocus sequence typing
MRSA	Methicillin-resistant *Staphylococcus aureus*
MTBC	*Mycobacterium tuberculosis* complex
NBS-LRR	Nucleotide-binding site, leucine rich repeat
NLR	Nucleotide oligomerization domain-like receptors

NTM	Non-tuberculous *Mycobacterium*
ORF	Open reading frame
PAMP	Pathogen-associated molecular pattern
PBP	Penicillin-binding protein
PCR	Polymerase chain reaction
PMU	Potential mobile unit
PRR	Pattern recognition receptor
RLR	Retinoic acid-inducible gene-1-like receptors
ROS	Reactive oxygen species
SADS	Swine acute diarrhoea syndrome
SARS	Severe acute respiratory syndrome
SIV	Simian immunodeficiency virus
ST	Sequence type
T3SS	Type 3 secretion system
T4SS	Type 4 secretion system
T6SS	Type 6 secretion system
TB	Tuberculosis
TLR	Toll-like receptor
UNICEF	United Nations International Children's Emergency Fund
UTI	Urinary tract infection
UV	Ultra-violet (light)
VOC	Volatile organic compound
WHO	World Health Organization
XDR	Extensive drug resistance

Part I
An Introduction to the Background Science

1

The Role of Nucleic Acid Analysis in Understanding Infectious Diseases

To understand the natural history of infectious diseases we first need to know the identity of the organisms with which we are dealing and their relationships with other organisms. The science of classification is called taxonomy and its objective is to arrange cellular organisms into groups or taxa on the basis of their mutual similarity or evolutionary relatedness. There are eight major taxonomic ranks (Figure 1.1) and the lower down the ranking that two or more organisms are the more closely related they are. Thus, organisms belonging to the same genus would be expected to have many properties in common. At the upper level of the hierarchy of ranks there are three domains: Archaea, Bacteria, and Eucarya. Microorganisms are found in all three domains but, to date, no members of the Archaea have been found to cause disease of any kind. Rather, the ability of cellular organisms to cause infectious disease is confined to Bacteria and some microbial members of the Eucarya. Viruses also cause disease but as they are not cellular organisms they are classified separately on the basis of their morphology, type of nucleic acid, and mode of replication.

Identification represents the practical side of taxonomy and is the process of determining if a particular isolate belongs to a recognized taxon. Identification of an organism causing a particular disease is essential because in some cases, infection with completely unrelated bacteria can result in similar clinical symptoms. For example, osteomyelitis is an infection of bone and can be caused by *Staphylococcus aureus*, *Streptococcus* species, *Enterobacter* species, *Haemophilus influenzae*, and occasionally fungi.

The classical method of identifying microorganisms was to determine their morphology by microscopy and their metabolic properties. Morphology is of limited use for bacterial identification as > 90% are rod shaped. Consequently, bacterial identification was based on physiological

Microbiology of Infectious Disease. Sandy B. Primrose, Oxford University Press.
© Sandy B. Primrose (2022). DOI: 10.1093/oso/9780192863843.003.0001

LIFE			
DOMAIN	Bacteria	Bacteria	Eucarya
KINGDOM			Fungi
PHYLUM	Firmicutes	Proteobacteria	Basidiomycota
CLASS	Bacilli	Gamma-proteobacteria	Tremellomycetes
ORDER	Bacillales	Enterobacterales	Tremellales
FAMILY	Staphylococcaceae	Enterobacteriaceae	Tremellaceae
GENUS	*Staphylococcus*	*Escherichia*	*Cryptococcus*
SPECIES	*aureus*	*coli*	*neoformans*

Figure 1.1. Examples of the eight major taxonomic ranks. Note that there is disagreement among taxonomists about the number of domains and kingdoms but this has no impact on the biology of microorganisms.

characteristics: ability to utilize various carbon and nitrogen sources, resistance to certain compounds, ability to grow in the absence of oxygen, etc. Over time, it was realized that this method had a significant failing: unrelated bacteria often were placed in the same family or even the same genus. This misclassification was highlighted with the development of methods for determining the percentage of guanine + cytosine (%GC) in a nucleic acid. The GC ratio of DNA from higher animals and plants is 30–50% with an average around 40%. By contrast, the %GC ratio of microbial nucleic acids varies from 20% to 80%. Many organisms that were believed to be related based on physiological properties were found to have significantly different %GC content and had to be reclassified.

The next significant advance in classification came with the development of sequence analysis of the genes encoding 16S ribosomal RNA (rRNA) of prokaryotes and the 18S rRNA of eukaryotes. Ribosomes are an essential part of the cellular machinery for synthesizing proteins and consequently rRNA genes are highly conserved. Thus, their sequences are believed to accurately represent evolutionary changes. The current phylogenetic (evolutionary) tree based on rRNA sequence analysis is shown in Figure 1.2. Whilst this tree shows the relationships between different groups of organisms at a coarse level, it is not very useful for investigating organisms in the same genus nor for determining their natural history. These kinds of studies require a knowledge of entire genome sequence of organisms.

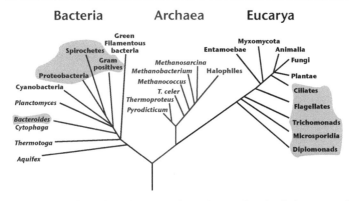

Figure 1.2. The tree of life based on rRNA analysis. The shaded areas indicate the taxa where pathogens are found although *Bacteroides* species are mostly opportunistic pathogens. The vast majority of Gram-negative bacteria can be found within the Proteobacteria (Figure 1.1).

Source: Modified from Eric Gaba/adaptation of NASA Astrobiology Institute website figure (Public domain). <https://commons.wikimedia.org/wiki/File:Phylogenetic_tree.svg>.

DNA Sequencing Technology

The properties of microbes are governed by their genomes and without genomic sequence data it was nearly impossible to understand their physiology fully, their real capabilities, or their evolution. DNA sequencing technology was developed in the late 1970s and the first microbial genome sequenced was that of bacteriophage ΦX174. Although this genome is only 5,386 bases in length, the assembly of the sequence data and its analysis took a number of people many months. It was nearly 20 years later, in 1995, before the first bacterial genomes were sequenced: *Haemophilus influenzae* (1,830,137 base pairs) and *Mycoplasma genitalium* (580,070 base pairs). As a measure of the difficulty of the early DNA sequencing methodology, it took 10 years and $2.7 billion to generate the first sequence of the human genome (~ 3 billion base pairs) which was published in 2003.

Starting in 2005, newer, high-throughput sequencing methodologies (next-generation sequencing) were developed and today it is possible to sequence a whole bacterial or viral genome in a matter of hours and the cost of doing so is less than $100. The latest instrumentation from Oxford Nanopore Technologies even allows sequencing to be done outside the laboratory and is invaluable for studying disease outbreaks in remote places. As a consequence of this new methodology, genome sequences for over 300,000 (!) microbial strains were available at the time of writing. When coupled with

---

Box 1.1 TRACKING NEW VARIANTS OF COVID-19

The virus causing the COVID-19 pandemic has an RNA genome and, unlike DNA replication which is subject to proofreading, RNA replication is prone to error. This means that with millions of people infected during the pandemic, a large number of virus variants were circulating globally. The key question for epidemiologists was whether any of these mutants were being positively selected and if they were, did they have increased infectivity and/or result in higher mortality? The only way to answer this question was to sequence large numbers of isolates and this is what COVID-19 Genomics UK (COG-UK) did. Practically, it was a group of ~ 400 scientists across more than thirty UK universities and institutions linked in a network. Nine months after the disease hit the United Kingdom, COG-UK had sequenced over 250,000 COVID-19 genomes and had identified four key variants: three in Britain and a fourth in Brazil. One of the British variants (D614G) had a change in amino acid 614 of the viral spike protein with glycine (G) replacing aspartate (D). Population genetic analysis indicated that 614G increased in frequency relative to 614D in a manner consistent with a selective advantage. However, there was no indication that patients infected with the spike 614G variant have higher COVID-19 mortality or clinical severity, but 614G was associated with higher viral load and younger age of patients.

---

sophisticated data analysis software, these sequence data can be mined for almost any kind of information that the microbiologist might want including identifying genes or putative genes (open reading frames, ORFs) that encode proteins, genes encoding RNA, regulatory sequences, structural motifs, and repetitive sequences. A comparison of genes within a species or between different species can show similarities between protein functions, or the relations between species. Armed with this information we now can get a detailed understanding of the natural history of microbial diseases and this is the theme of this book. A good example of the use of this technology was the large-scale sequencing of isolates of the virus that caused the COVID-19 pandemic (Box 1.1).

Using DNA Sequencing to Characterize Microbial Strains

Multilocus sequence typing (MLST) is a technique in molecular biology for the typing of multiple loci. The procedure characterizes isolates of microbial species using the DNA sequences of internal fragments of seven housekeeping genes, that is, genes that encode essential functions and are present in all isolates of a particular organism. Approximately 450–500 base pair (bp) internal fragments of each gene are used as these can be accurately sequenced on both strands using an automated DNA sequencer. For each housekeeping gene, the different sequences present within a bacterial species are assigned as distinct alleles and, for each isolate, the alleles at each of the loci define the allelic profile or sequence type (ST). As there are many alleles for each of the seven genes, isolates are unlikely to have identical profiles by chance. Thus, isolates with the same allelic profile can be assigned as members of the same clone.

Because classical MLST relies on the sequencing of only seven housekeeping genes it provides only a moderate typing resolution. Since next-generation sequencing enables high-throughput analyses of entire bacterial genomes at an affordable cost, it has quickly become indispensable for performing population and outbreak analyses. However, appropriate bioinformatics tools are necessary as a prerequisite for handling and interpreting sequence data. Core genome MLST (cgMLST) aims to combine the discriminatory power of classical MLST with the extensive genetic data derived from whole genome sequencing. By exploiting hundreds of gene targets of the entire bacterial genome, cgMLST provides maximum resolution for surveillance analyses.

Analysing Ancient DNA (aDNA)

To understand the evolution of a disease it is necessary to look at the changes in the genome of a pathogen over time. This means recovering and analysing DNA from victims who died 100 or more years ago. However, such ancient DNA often is of very poor quality compared with DNA isolated from fresh tissue. On storage, DNA undergoes fragmentation, cross-linking, and deamination and the extent of these changes depends on the storage temperature and the chemical composition of the environment surrounding the DNA. For this reason, recovering aDNA is fraught with problems and initial studies were restricted to pathogens causing diseases that can be identified by skeletal lesions as this increased the chance of successful extraction. The first successes came with the isolation of *Mycobacterium tuberculosis* and *Mycobacterium leprae* DNA from ancient bones. Subsequently, aDNA from *Mycobacterium tuberculosis* was recovered from mummies as the mummification process prevents DNA degradation. This enabled the identification of tuberculosis in a mummified specimen from the Andes that was over 1,000 years old. Over time, the range of suitable specimens for sourcing aDNA extended to calcified pleura, dental pulp, coprolites (mummified or petrified faeces), and museum specimens of paraffin-embedded sections (Figure 1.3).

The early studies on aDNA used the polymerase chain reaction (PCR) to amplify short genomic segments recovered from ancient specimens. Consequently, such studies could be used to confirm the presence of pathogen DNA but gave little information about its evolutionary history. The development of two technologies made it possible to reconstruct whole pathogen genomes from ancient materials: next-generation sequencing and targeted enrichment (Figure 1.3). In the latter, DNA or RNA from a modern isolate of the target organism is used as a bait for the aDNA, allowing it to be freed of contaminating DNA before it is amplified and sequenced. Targeted enrichment led to the recovery of the first ancient bacterial genome from the teeth of a fourteenth-century plague victim (p27).

Dating Evolutionary Events: Molecular Clocks

The principle of the molecular clock is that DNA and protein sequences evolve at a rate that is relatively constant over time for a particular group of organisms. This rate is the product of the number of mutations that arise per replication event, the frequency of replication events per unit time, and the probability of mutational fixation. A direct consequence of this constancy is that the genetic difference between any two species is proportional to the time since these species last shared a common ancestor. For this approach to be useful it is necessary to calibrate the molecular clock for the particular organisms being studied because different groups of organisms evolve at different rates. For example, in *Neisseria gonorrhoeae* the rate is 10 base substitutions per megabase (1 million nucleotides) per year but in *Mycobacterium tuberculosis* it is only 0.001. Also, RNA viruses mutate at a much higher rate than DNA viruses because RNA replication is error prone. In practice, calibrating the molecular clock is a lot more difficult than it sounds because it is necessary to have isolates that have been collected at a particular time and remained uncultured since then. A good example of a suitable sample is the recovery of DNA of the bacterium *Pasteurella pestis* from the pulp of teeth of plague victims discussed in the previous section. The age of the victims can be determined either by carbon dating or from burial records. However, for most pathogens, getting suitable samples can be very difficult and this is particularly true for viruses.

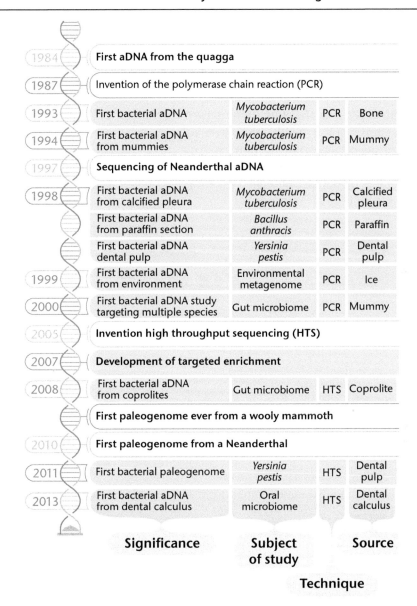

Figure 1.3. The history of bacterial aDNA research. Breakthroughs in bacterial studies are highlighted in blue. PCR, polymerase chain reaction; HTS, high throughput sequencing. Figure modified from Arning and Wilson (2020) with permission.

Horizontal Gene Transfer

Horizontal gene transfer, sometimes called lateral gene transfer, is the movement of genetic material between unicellular and/or multicellular organisms other than by the ('vertical') transmission of DNA from parent to offspring (reproduction). It is the most important factor in the evolution of many microorganisms. It is the primary mechanism for the spread of antibiotic resistance in bacteria (Box 1.2) and has played a major role in the evolution, maintenance, and transmission of virulence. The principal methods of horizontal gene transfer are transformation, conjugation, and transduction (Figure 1.4).

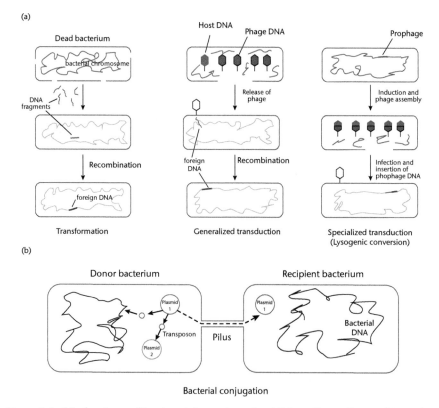

Figure 1.4. Mechanisms of horizontal gene transfer. (A) transformation and transduction; (B) conjugation.

Transformation is an active process whereby free DNA is taken up by an organism. Cells which can undergo genetic transformation are said to be competent. Some bacteria such as *Acinetobacter* and the pneumococcus are permanently competent but in other bacteria, such as *Vibrio cholerae*, competence is induced by environmental stimuli. Many bacteria, such as *Escherichia coli*, are not known to be naturally competent but competence can be achieved artificially in the laboratory. Once the extraneous DNA has entered a new cell, it needs to be integrated into the host chromosome. The reason for this is that DNA that does not have an origin of DNA replication (*ori*) will be lost during successive bacterial divisions. Usually, a piece of the new DNA will *replace* a corresponding piece of host DNA by homologous recombination but in rare cases it may be *added* to the host genome by insertion, a process known as illegitimate recombination.

Conjugation is a process whereby DNA is transferred from one cell (donor or 'male') to another (recipient or 'female') by means of a tubular structure known as a pilus. Many cells contain extrachromosomal DNA elements known as plasmids that are able to replicate independently of the host chromosome because they have an *ori* sequence. Some plasmids encode genes that enable the plasmid to be transferred to new cells and such plasmids are said to be conjugative or self-transmissible. Some conjugative plasmids can only be transferred between, and maintained in, closely related bacteria but others are much more promiscuous and can spread to unrelated bacteria. Other plasmids are non-conjugative and transmission to new cells occurs rarely unless a conjugative plasmid also is present in the donor cell. Many plasmids contain genes associated with pathogenicity or antibiotic resistance (Box 1.2) and the latter are particularly worrying as they are making it difficult to treat people with infections successfully.

Transduction is the process whereby DNA transfer is mediated by a bacteriophage (bacterial virus, or 'phage'). When a phage infects a bacterial cell,

---

Box 1.2 THE SPREAD OF ANTIBIOTIC RESISTANCE

Plasmids often carry multiple antibiotic resistance genes, contributing to the spread of multidrug-resistance (MDR). Multiple resistance genes are commonly arranged in resistance cassettes. The antibiotic resistance genes found on the plasmids confer resistance to most of the antibiotic classes used nowadays, for example, beta-lactams, fluoroquinolones and aminoglycosides. It is very common for the resistance genes or entire resistance cassettes to be re-arranged on the same plasmid or be moved to a different plasmid or chromosome because they are part of transposons.

Antibiotic resistance mediated by MDR plasmids severely limits the treatment options for the infections. The global spread of MDR plasmids has been enhanced by selective pressure from inappropriate antibiotic usage in human and veterinary medicine coupled with the fact that most of the resistance plasmids are conjugative or can be mobilized by a conjugative plasmid. Conjugative transfer normally occurs between related bacteria but on rare occasions can take place between genetically unrelated bacteria. For the latter to occur, the plasmid must be able to replicate in its new host and the antibiotic resistance genes must be able to be expressed there as well.

only its nucleic acid gets into the cytoplasm. This nucleic acid undergoes replication and then directs the synthesis of viral proteins. Late in infection, the viral proteins encapsulate the viral nucleic acid and the newly assembled virus particles are released from the cell by lysis. Occasionally, a mistake occurs during virus assembly and fragments of host cell DNA get encapsulated instead of viral DNA. When such viruses infect a new cell, the DNA fragments from the previous host are released. As with transformation, these fragments need to be incorporated into the new host's genome. Any part of the genome can be packaged by the viral protein and the phages in which this has occurred are known as generalized transducing phages.

Some bacterial viruses, known as temperate phages, have the ability to insert their DNA into the genome of a host cell as an alternative to replication and lysis. This non-lytic process is known as lysogeny and the inserted viral genome is known as a prophage. When the phage begins its lytic cycle, occasional errors may occur whereby some non-essential phage DNA is replaced with a piece of adjacent bacterial DNA. The phages carrying the bacterial DNA still are able to replicate and carry the bacterial DNA with them as a permanent feature. If the bacterial DNA encodes a gene or genes that can be expressed in a cell that has been lysogenized then that cell will acquire new properties, a process known as lysogenic conversion. There are many bacterial diseases, such as diphtheria, and cholera (p53), where phage-borne toxin genes are responsible for the pathogenicity of the host bacteria.

Mobile genetic elements (MGEs) are DNA sequences that can move from place to place in a particular genome or from cell to cell. They can be divided into two major groups: intercellular and intracellular. Examples of the first group are plasmids and bacteriophages, as described earlier, that are transmissible from cell to cell. Another example is the class of MGEs known as integrative conjugative elements (ICEs). These encode the machinery required for their own integration and excision from the bacterial genome and transfer between cells. The intracellular MGEs can move about within a cell but cannot move between cells unless they are integrated into a plasmid or bacteriophage. There are two types of intracellular MGE: insertion sequences and transposons.

Insertion sequences (IS) are short stretches of DNA, usually less than 1,000 base pairs that contain a gene encoding a transposase bounded by short, inverted repeat sequences. The transposase enables the IS to move around the genome and insert itself at random. If the point of insertion is in a gene or regulatory element then inactivation (mutation) is the most likely outcome. Some bacteria, such as *Shigella* species (p37), contain many IS elements that have contributed to increased pathogenicity. Transposons are similar to IS but they are larger and contain multiple genes in addition to the one encoding the transposase. Transposons can jump in and out of plasmids and play a key role in plasmid evolution and plasticity. In addition, both IS and transposons have played a key role in genome evolution.

The term *genomic island* is applied to a part of a genome that has evidence of having originated by horizontal transfer. A genomic island can encode many functions. If it contains many antibiotic resistance genes it is known as an antibiotic resistance island and if it has genes associated with pathogenesis it is known as a *pathogenicity island*. Many genomic islands are flanked by repeat structures and carry fragments of other mobile elements such as phages and plasmids. Some genomic islands can excise themselves spontaneously from the chromosome and insert into plasmid DNA thereby facilitating their transfer to new hosts.

The extent of horizontal gene transfer and its role in shaping the evolution of microorganisms has only become apparent in the last twenty or so years with the large-scale sequencing of many genomes. Horizontal gene transfer leaves traces of its occurrence (signatures) and genome assembly and analysis software can detect these. For example, genomic islands are usually detected because of their frequent association with tRNA-encoding genes and a different G+C content compared with the rest of the genome. Other compositional signatures are segments of DNA with different dinucleotide frequencies or changes in codon use. Genes transferred by specialized transduction can be identified if they are near phage-related sequences, such as phage integrases, or if they are near transfer RNAs which are preferential integration sites for many temperate phages. Accordingly, the presence of prophage sequences nearby putatively transferred genes could be taken as corroborating evidence of specialized transduction. Methods for identifying prophage sequences in isolate genomes typically rely on sequence similarity with known viral genes as well as identifying regions with viral genome characteristics such as shorter protein lengths and shared transcription directionality of adjacent genes.

Suggested Reading

Arning N. and Wilson D.J. (2020) The past, present and future of ancient bacterial DNA. *Microbial Genomics* **Jul** 6(7) doi:10.1099/mgen.0.000384

Baker S., Thomson N., Weill F-X., and Holt K.E. (2018) Genomic insights into the emergence and spread of antimicrobial-resistant bacterial pathogens. *Science* **360**, 733–8.

Duchêne S., Holt K.E., Weill F-X., Le Hello S., Hawkey J., et al. (2016) Genome-scale rates of evolutionary change in bacteria. *Microbial Genomics* 2(11) doi: 10.1099/mgen.0.000094

Hall J.P.J., Brockhurst M.A., and Harrison E. (2017) Sampling the mobile gene pool: innovation via horizontal gene transfer in bacteria. *Philosophical Transactions of the Royal Society B* **372**, 20160424

Land M., Hauser L., Jun S-R., Nookaew I., Leuze M.R., et al. (2015) Insights from 20 years of bacterial genome sequencing. *Functional & Integrative Genomics* **15**, 141–61.

Oxford Nanopore Technologies (2020) Large insights into microorganisms. www.nanoporetech.com/publications

Siguier P., Gourbeyre E., and Chandler M. (2014) Bacterial insertion sequences: their genomic impact and diversity. *FEMS Microbiology Reviews* **38**, 865–91.

Spyrou M.A., Bos K.I., Herbig A., and Krause J. (2019) Ancient pathogen genomics as an emerging tool for infectious disease research. *Nature Reviews Genetics* **20**, 323–40.

Tagini F. and Greub G. (2017) Bacterial genome sequencing in clinical microbiology: a pathogen-oriented review. *European Journal of Clinical Microbiology and Infectious Disease* **36**, 2007–20.

2

Some Common Factors Involved in Host-Pathogen Interactions

Many of the pathogens discussed in this book have evolved common mechanisms for effecting the early stages of infecting target organisms, be they plants or animals. These include secretion systems, iron-chelating systems (siderophores), and motility. The target organisms also have evolved common mechanisms for recognizing when they are being attacked, such as the detection of pathogen-associated molecular patterns by pattern recognition receptors. As might be expected, the pathogens fight back by mutating in various ways. This is no more evident than in the largely unsuccessful attempts by plant breeders to generate plants with long-lasting resistance to late potato blight caused by *Phytophthora infestans* (Chapter 23).

It is worth noting that not all pathogens need to come into direct contact with host cells. Certain *Clostridium* species, such as *Clostridium botulinum* (botulism) and *Clostridium tetani* (tetanus) produce protein toxins that on their own have devastating effects on mammalian cells.

Secretion Systems

Secretion systems are the protein complexes used by bacteria to transport substances, particularly proteins, across cell membranes. In the context of this book, they are the cellular devices whereby pathogenic bacteria secrete virulence factors that they use either to invade host cells or to attack cells biochemically. There are several different classes of bacterial secretion systems and their designs can differ based on whether their protein substrates cross a single phospholipid membrane (pathogens without cell walls), two membranes (Gram-positive bacteria), or even three membranes (Gram-negative bacteria) where two are bacterial and one is a host membrane. There are at least eight secretion systems specific to Gram-negative bacteria, four are specific to Gram-positive bacteria, and two are common to both. The most

Microbiology of Infectious Disease. Sandy B. Primrose, Oxford University Press.
© Sandy B. Primrose (2022). DOI: 10.1093/oso/9780192863843.003.0002

relevant in terms of the pathogens described in subsequent chapters are the type III, type IV, and type VI secretion systems. These secretion systems operate over and above the general export systems known as Sec and Tat. With the Sec system, unfolded proteins are recognized by the presence of a signal sequence at the N-terminus and during transport through the cell membrane to the periplasm, this signal sequence is removed. The Tat system is similar to Sec in the process of protein secretion but exports proteins only in their folded (tertiary) state.

The type III secretion system (T3SS) is a needle and syringe-like system (Figure 2.1) that injects virulence factors, usually proteins, directly into eukaryotic cells. T3SS substrates are generically called effector proteins. Pathogens may secrete only a few effector proteins, as in the cases of *Pseudomonas* (p82) and *Yersinia* (p29), or several dozen, as in the cases of *Shigella* (p34) and enterohaemorrhagic *Escherichia coli* (EHEC, p34).

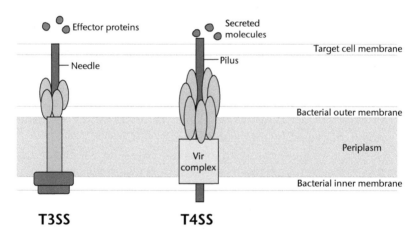

Figure 2.1. Schematic of type III and type IV secretion systems.

The type IV secretion system (T4SS) is similar to the pilus whereby bacteria that are participating in conjugation can unidirectionally transfer DNA (Figure 2.1). The T4SS can transport single proteins, protein–protein complexes, and DNA–protein complexes. Examples of bacterial pathogens that employ T4SSs for virulence are *Neisseria gonorrhoeae* (p69), which uses its T4SS to mediate DNA uptake (which promotes virulence gene acquisition), and *Legionella pneumophila* (p63) and *Helicobacter pylori* (p80), which use their T4SSs to translocate effector proteins into host cells during infection to disrupt their defence strategies.

The type VI secretion system (T6SS) is thought to resemble an inverted bacteriophage extending outward from the bacterial cell surface and consists of three sub-complexes: a phage tail-like tubule, a phage baseplate-like structure, and cell envelope-spanning membrane complex (Figure 2.2). The phage tail-like component of the T6SS is a dynamic tubular structure that undergoes cycles of assembly and disassembly. The tubules consist of repeating units of the proteins VipA and VipB arranged as a sheath around a tube (Figure 2.3) built from stacked hexameric rings of the haemolysin co-regulated protein (Hcp). At the tip of the Hcp tube sits a trimer of the phage tail spike-like protein, VgrG. Contraction of the sheath is thought to propel the Hcp tube, VgrG, and associated substrates through the membrane of a neighbouring cell. The tubule structure is dismantled through the action of the protein ClpV, which sits at the tubule base.

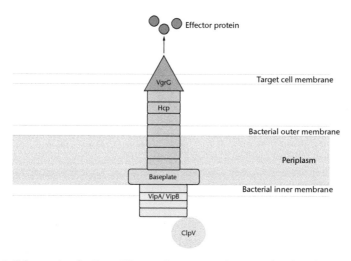

Figure 2.2. Schematic of a Type VI secretion system (see text for details).

Figure 2.3. The tubular section of a Type VI secretion system.

Source: Astrojan/Wikimedia Commons originating from <http://www.rcsb.org/ pdb/explore/explore.do?structureId= 3wx6>. Reproduced under Creative Commons Attribution 4.0 International licence (CC BY 4.0).

Type VI secretion systems translocate proteins into a variety of recipient cells, including eukaryotic cell targets and other bacteria. In addition to their role in pathogenicity, T6SSs are used by bacteria to defend themselves against eukaryotic predators. Notable pathogens which have a T6SS are *Pseudomonas aeruginosa* (p82) and *Vibrio cholerae* (p56).

Pathogen-Associated Molecular Patterns

The innate immune system of higher animals constitutes the first line of host defence during infection and as such, plays a crucial role in triggering a proinflammatory response to invading pathogens. It relies on the recognition of evolutionary conserved structures on pathogens, known as pathogen-associated molecular patterns (PAMPs), by pattern recognition receptors (PRRs) on the target cell. Different components of bacterial cells can act as PAMPs (Figure 2.4) including cell wall components and flagella. In the case of fungi, the major PAMPs are components of the cell wall: β-glucans, chitin (β-1,4-linked homopolymer of *N*-acetylglucosamine) and mannans (β-1,4-linked homopolymer of mannose). The major PAMP of protozoan parasites is glycophosphatidylinositol (GPI), a glycolipid that can be attached to the C-terminus of proteins in the cell membrane.

In higher animals, the best known PRRs are the toll-like receptors (TLRs) and they can be divided into subfamilies that recognize related PAMPs. TLR1,

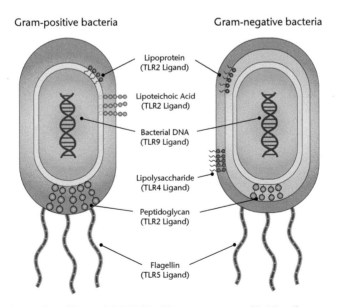

Figure 2.4. Examples of bacterial PAMPs. Figure courtesy of InVivoGen.

TLR2, TLR4, and TLR6 recognize lipids whereas TLR3, TLR7, TLR8, and TLR9 recognize nucleic acids. Upon PAMP recognition, PRRs signal to the host the presence of infection and trigger the synthesis and/or activation of a broad range of defensive molecules including cytokines, chemokines, and immunoreceptors that play key roles in innate immunity. In addition to TLRs, there is a large group of other cytosolic PRRs knowns as RLRs and NLRs. The presence of nucleic acids in the cytoplasm of cells as a result of the un-coating of viruses is recognized by the innate immune system independently of TLRs, RLRs, and NLRs.

Plants also have PRRs but these operate in a different way to those of higher animals. Bacterial plant pathogens inject effectors into target cells using a type III secretion system. Fungi and Oömycetes, on the other hand, invaginate a feeding organelle called a haustorium into a host cell. This angioplasty balloon-like structure sets up an interface through which effectors travel by an as-yet undefined mechanism. Despite differences in the mechanism of effector delivery, the end result is the same: suites of pathogen proteins are introduced into plant cells where they contribute to virulence.

Plants have evolved specific *R* protein alleles to detect specific plant pathogen effectors. Detection of an effector by an *R* protein triggers rapid activation of a very effective defence. The majority of plant *R* genes encode NBS-LRR proteins (nucleotide-binding site, leucine-rich repeat) and comparative genomics has demonstrated that all plants maintain large collections of these genes, many of which may have alternate alleles. Although each individual NBS-LRR allele typically provides resistance to a single pathogen isolate, the collective surveillance capability of a plant's NBS-LRRs appears quite broad. If NBS-LLR proteins bind pathogen effectors it activates effector-triggered immunity (ETI). This response results in hypersensitivity which is a form of programmed cell death that limits pathogen growth and effects plant resistance. Failure to bind effectors results in effector-triggered susceptibility (ETS) and leads to plant disease. This topic is discussed in more detail in Chapter 23.

Siderophores and Lipocalin 2

Iron is an essential element for virtually all living organisms except lactic acid bacteria and is used to catalyse a wide variety of indispensable enzymatic reactions. In oxygen-rich conditions, iron exists as the insoluble ferric ion and in higher animals it is maintained in a soluble form by binding to iron-chelating proteins such as transferrin and lactoferrin. These proteins have a very high association constant for ferric iron and normally are only about 30–40% saturated resulting in almost negligible concentrations

$(10^{-24}$ mol L$^{-1})$ of free iron. This chelation of iron presents pathogenic bacteria and fungi with a challenge: how to acquire sufficient iron during infection. They solve this problem by secreting high-affinity iron-chelating compounds known as siderophores. Over 250 different siderophores have been identified and this wide variety may be due to evolutionary pressures placed on microbes to produce structurally different siderophores which cannot be transported by other microbes' specific active transport systems.

Siderophores have such high affinity for iron that they can strip the metal from lactoferrin and other iron-binding host proteins. In response, host cells secrete lipocalin 2 which is part of the innate immune system. This protein is stored in neutrophils and is rapidly secreted when infection is detected by recognition of PAMPs. Infection also induces the synthesis of high levels of lipocalin 2. This lipocalin 2 binds to siderophores and prevents their uptake by the infecting microorganisms thereby starving the pathogen of iron. Some pathogens have evolved siderophores that cannot be bound by lipocalin 2. For example, the siderophore enterobactin that is produced by *Escherichia coli* is readily bound by lipocalin 2 but the glycosylated enterobactin (known as salmochelin) produced by *Salmonella* species is not.

Different microorganisms have to compete with one another for nutritional resources in most environmental niches. Some bacteria, such as actinomycetes, give themselves a competitive advantage by secreting compounds called sideromycins or microcins that are siderophores conjugated to antimicrobial compounds. The siderophore effectively is a Trojan horse that exploits the siderophore uptake pathways to get the antimicrobial compound into the target cell. Pharmaceutical companies have begun exploiting this phenomenon in the search for new antibiotics. Rather than screening large numbers of actinomycetes for compounds of interest, they have been conjugating existing antibiotics to siderophores. The first of these synthetic antibiotics, cefiderocol, is a cephalosporin conjugated to a siderophore. It received regulatory approval in 2020 for the treatment of infections due to multi-drug-resistant Gram-negative bacteria.

Bacterial Motility and Chemotaxis

Except in the case of wound infections, pathogenic bacteria usually interact with epithelial surfaces. Most internal epithelial surfaces are covered with a hydrogel layer that is a few hundred micrometres in thickness. This hydrogel consists of cross-linked mucin molecules and is the first line of defence against bacterial invasion. If bacteria cannot come in direct contact with the epithelial surface then they cannot inject effectors into the target cell. A pathogen needs two properties in order to penetrate the hydrogel: motility and chemotaxis. Most motile bacteria have multiple flagella and if all of these rotate in the same direction the bacteria are propelled forward ('running'). If one or more flagella rotate in the opposite direction then the bacteria will switch direction ('tumbling'). This combination of running and tumbling is known as a random walk and enables the bacteria to move about 300 microns within a few minutes. However, the viscosity of the hydrogel biases the random walk such that the bacteria spread over the hydrogel rather than through it.

Chemotaxis is a behaviour whereby bacteria detect external chemical changes along their swimming trajectory and swim up concentration gradients by means of a biased random walk. That is, runs up the concentration gradient are longer than runs down the gradient. Pathogenic bacteria sense chemicals secreted by epithelial cells that are diffusing through the hydrogel and stimulates them to penetrate the hydrogel. Non-chemotactic mutants of pathogenic bacteria have greatly reduced or no pathogenicity indicating the importance of chemotaxis. It is believed that bacteria that have a spiral morphology, for example, *Treponema pallidum* that cause syphilis (p90) and *Helicobacter pylori* that causes gastric ulcers (p76), can bore their way through the hydrogel much easier than rod-shaped bacteria.

Phagolysosomes

A key defence strategy of higher animals is the engulfment of pathogens by phagocytes such as macrophages, neutrophils, and dendritic cells. During phagocytosis, a phagosome is formed by the fusion of the phagocyte cell membrane around the invading microorganism (Figure 2.5). The phagosomes then fuse with lysosomes to form phagolysosomes. The lysosomes contain hydrolytic enzymes and reactive oxygen species (ROS) which kill and digest the pathogens. Neutrophils contain myeloperoxidase which produces toxic oxygen and chlorine derivatives to kill pathogens in an oxidative burst. They also release proteases and anti-microbial peptides into the phagolysosome. Macrophages lack myeloperoxidase and rely on phagolysosome acidification, glycosidases, and proteases to digest microbes. Any useful materials from the digested particles, such as amino acids, are moved into the cytosol, and unwanted materials are removed by exocytosis.

Some bacteria exploit phagocytosis as an invasion strategy. They either reproduce inside of the phagolysosome (e.g. *Legionella pneumophila*, p63) or escape into the cytoplasm before the phagosome fuses with the lysosome (e.g. *Rickettsia* spp.). Many mycobacteria, including *Mycobacterium tuberculosis,* can manipulate the host macrophages to prevent lysosomes from fusing with phagosomes and creating mature phagolysosomes.

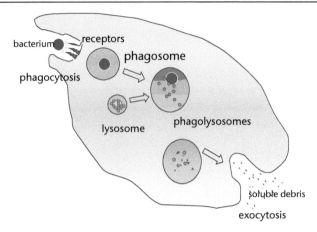

Figure 2.5. The process of phagocytosis.
Source: Graham Colm/Wikimedia Commons. Reproduced under Creative Commons Attribution 3.0 Unported license. <https://commons.wikimedia.org/wiki/File: Phagocytosis2.png>.

Such incomplete maturation of the phagosome maintains an environment favourable to the pathogens inside it.

Suggested Reading

Amarante-Mendes G.P., Adjemian S., Branco L.M., Zanetti L.C., Weinlich R., and Bortoluci K.R. (2018) Pattern recognition receptors and the host cell death machinery. *Frontiers in Immunology* **9** doi:10.3389/fimmu.2018.02379

Green E.R. and Mescas J (2016) Bacterial secretion systems: an overview. *Microbiology Spectrum* **4**(1) doi:10.1128/microbiolspec.VMBF-0012-2015

Kramer J., Özkaya Ö., and Kümmerli R. (2020) Bacterial siderophores in community and host interactions. *Nature Reviews in Microbiology* **18**, 152–63.

Part II
Bacterial Pathogens

3

The Three Great Pandemics of Plague

It is rare for an outbreak of an infectious disease to be sufficiently important to be described in school history books. The Black Death in Europe is one such disease, probably because of its huge social and economic impacts. The outbreak began in 1347 after a Crimean Tatar force attacked Kaffa, a Genoese trading port on the Black Sea. The attack was unsuccessful because there was an outbreak of plague among the Tatar troops. Before they withdrew, the Tatars engaged in biological warfare by catapulting infected corpses into the besieged city. Infected Genoese sailors subsequently sailed from Kaffa to various Mediterranean ports thereby introducing the Black Death into Europe.

The plague reached England in June 1348 when an infected seaman landed in Weymouth, Dorset from south-west France. From Weymouth the disease spread west to Bristol and north to London. By March 1349 the plague was well established in southern England when a ship carrying the disease arrived in Hull and introduced it to northern England. By the end of 1349 the plague epidemic was beginning to die out but over 30% of the population of England had died. In France, the death rate was nearer 50%. One consequence of the Black Death was an extended pause in the 100 Years War between England and France. Another was a shortage of manpower which led to changes in the relationship between labourers and landowners in England.

Although the first episode of the Black Death died out in Europe around 1351, there were repeated occurrences over the next 400 years, although not with the severity of the original outbreak. One city that tried to control the outbreaks was the port of Venice. In 1403 they introduced a law requiring travellers from infected areas to be isolated in a hospital for forty days, the *quarantena* or quarantine, to determine if they became ill and died or remained healthy. Later, in 1776, the authorities established a quarantine station on the island of Poveglia in the Venetian lagoon.

Microbiology of Infectious Disease. Sandy B. Primrose, Oxford University Press.
© Sandy B. Primrose (2022). DOI: 10.1093/oso/9780192863843.003.0003

In most cases of plague, the first signs are large purulent abscesses in the groin known as buboes (Figure 3.1), hence the name bubonic plague. The disease often progressed to bacteraemia and this could lead to death from septic shock. Alternatively, the disease could spread through the blood to the lungs resulting in a fatal pneumonia. Pneumonic plague, as this condition is called, is particularly important because it can be spread from person to person in aerosols. A major outbreak of pneumonic plague occurred in Europe from 1665 to 1666 which Samuel Pepys dubbed the Great Plague of London. Many of the victims of the 1665 outbreak in London were buried in the Bedlam burial ground close to what today is Liverpool Street Station. During the construction of the new Crossrail line (2009–2019), some of the burial pits had to be excavated. To the surprise of archaeologists from the Museum of London Archaeology, the burial practices appeared to be very neat rather than chaotic, with all the bodies laid out carefully and in an orderly fashion.

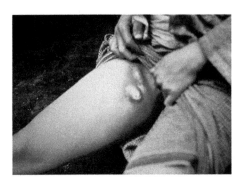

Figure 3.1. A plague patient displaying a swollen, ruptured inguinal lymph node, or buboe.

Source: Public domain, Centers for Disease Control and Prevention's Public Health Image Library (PHIL), ID #2047, via Wikimedia Commons. <https://commons.wikimedia.org/wiki/File:Plague_-buboes.jpg>.

Fortunately, the Great Plague of 1665 was the last major outbreak of plague in Britain. However, in 1894 a new pandemic of plague broke out in China and Hong Kong and from there spread to India and then to many parts of the world. Over the next sixty-five years there were sporadic outbreaks throughout the world before the pandemic ended in 1959. In 1894, when the outbreak started, Franco-Swiss physician Alexandre Yersin was working in Vietnam for the French colonial health service. At the request of the Pasteur Institute in Paris, he moved to Hong Kong to try to isolate the organism causing the plague. Soon he isolated a bacterium from the buboes of victims and showed that it killed mice. Today, this bacterium is known as *Yersinia pestis*. Yersin noted that the streets of Hong Kong were littered with dead rats and he found that they were infected with the same bacterium that he had isolated from victims of the plague. He suspected that there was a link between rats and the plague in humans.

In 1898, another French doctor called Paul-Louis Simond was working in India with plague patients. He noted that the legs and feet of patients were covered in blisters that were filled with plague bacteria. He thought

Box 3.1 FLEA TRANSMISSION OF PLAGUE

Yersinia pestis, the bacterial cause of plague, has a flea-borne transmission life cycle (Figure 3.2) involving rodents and the different fleas that parasitize them. If the fleas feed on an infected host they have the ability to transmit the bacterium if they have their next blood meal on a naïve host. This mode of transmission, termed early-phase transmission, is not seen unless five or more infected fleas feed simultaneously on the naïve host. A second phase of transmission can occur after *Yersinia pestis* grows as a biofilm in the valve between the oesophagus and midgut of the flea. This biofilm eventually blocks the incoming flow of blood when the flea attempts to feed resulting in regurgitative transmission of bacteria into the bite. Complete blockage can take up to one to two weeks or longer to develop, depending on the number of bacteria ingested in the infectious blood meal and the feeding frequency of the flea. Studies have shown that a higher percentage of transmissions by blocked fleas lead to death whereas early-phase transmissions more often lead to an immune response by the host and survival.

Figure 3.2. An Oriental rat flea (*Xenopsylla cheopis*) infected with the plague bacterium (*Yersinia pestis*), which appears as a dark mass in the gut.

Source: National Institute of Allergies and Infectious Diseases (Public domain). <https://commons.wikimedia.org/wiki/File:Flea_infected_with_yersinia_pestis.jpg>.

that these blisters preceded the formation of buboes and that they might have been caused by the bite of fleas. Subsequently, he showed that fleas transmitted the plague from infected rats to healthy rats (see Box 3.1). When an infected rat dies from the plague, the fleas on it immediately leave and find another warm-blooded host, thereby transmitting the disease. Later, in 1902, there was an outbreak of plague in Sydney, Australia. The chief medical officer, John Ashburton-Thompson, noted that an epizootic in rats always preceded an outbreak of plague in humans and that during the epizootic the rats carried many more fleas than normal.

A major outbreak of plague, known as the Plague of Justinian, occurred in the Mediterranean area between 541 and 543 and recurred about eighteen times over the next eighteen years. There are fewer skeletal remains of these victims but, the extraction of *Yersinia* DNA from teeth confirmed the cause of death. So, now we have conclusive evidence of three great plague pandemics. The first began with the Justinian Plague (541–549) and continued intermittently until 750. The second began with the Black Death in Europe (1347–1351) and there were successive waves, including the Great Plague (1665–1666), until the eighteenth century. Finally, the third pandemic began in China in 1894 and there were sporadic episodes until the mid-twentieth century. Many of these episodes were a result of infected rats being spread worldwide by steamships. The bacteria from the latter today are maintained enzootically in rodent populations in eastern Europe, Asia, Africa, and the Americas and occasionally cause local outbreaks for reasons that are not understood. The most recent of these was in Madagascar in 2017. However, death from the plague occurred long before the First Pandemic of 541–543: *Yersinia* DNA has been isolated from the teeth of human remains from the Bronze Age that are 2,800–5,000 years old.

Yersinia pseudotuberculosis is a zoonosis which is widespread in the environment and is usually transmitted to humans through consumption of contaminated food, particularly vegetables. The disease that it causes, Far East scarlet fever, is seldom fatal and resolves itself in one to three weeks. Molecular clock analyses show that *Yersinia pestis* and *Yersinia pseudotuberculosis* had a common ancestor about 5,000 to 6,000 years ago. Subsequent to this common ancestor, what genetic changes occurred to make *Yersinia pestis* one of the deadliest bacteria encountered by humans? For a start, *Yersinia pestis* acquired two new plasmids during its divergence from *Yersinia pseudotuberculosis* and both contain genes that contributed to evolution of the flea-borne transmission route. The pMT1 plasmid encodes the *ymt* gene and the pPCP1 plasmid encodes the *pla* gene. Expression of the latter is not required in the flea but is essential for bacterial invasiveness from the flea bite site following transmission.

Transmission of *Yersinia pestis* by fleas requires the bacterium to carry the gene for Yersinia murine toxin (*ymt*). This gene encodes a phospholipase that enables the bacterium to survive in the flea gut and to grow there to high titres. Flea transmission also is dependent on loss of function mutations in certain genes of the bacterium. Analysis of strains of *Yersinia pestis* taken from skeletons of different ages suggest that these mutations, and the acquisition of *ymt*, occurred around 3700 BC. As some of the Bronze Age bacterial isolates lacked these genetic changes, they probably were not transmitted by fleas but in aerosols from patients with pneumonic plague.

A key pathogenicity determinant of *Yersinia pestis* is the plasminogen activator (*pla*) gene. The protein that it encodes activates host fibrinolysis and allows the spread of the bacteria to the lymph nodes as well as their extensive growth in the lower airways. The presence of the *pla* gene is essential for the development of both bubonic and pneumonic plague. This gene was acquired at some time after the divergence of *Yersinia pestis* and *Yersinia pseudotuberculosis* but the protein that it encoded had isoleucine as amino acid 258. This means that the organism could only have caused pneumonic plague. Around 1000 BC a mutation occurred in the *pla* gene such that protein now had valine at position 258 and this enabled the bacterium to cause bubonic plague. Interestingly, some strains isolated from seventeenth-century victims lack the *pla* gene and the spread of these strains may have contributed to the disappearance of the second plague pandemic in eighteenth-century Europe.

Yersinia pestis outer membrane proteins (Yops) constitute an array of pathogenicity effectors that are translocated into host cells by a type III secretion system which is encoded on yet another virulence plasmid. One of these Yops, YopM, inhibits the release of proinflammatory cytokines by the infected host. Interestingly, humans carrying mutations that result in familial Mediterranean fever (FMF) have increased resistance to plague because they have enhanced release of the cytokine IL-1β. The FMF mutations appear to have arisen around 1,800 years ago, just before the Justinian Plague, in groups of people originating from around the Mediterranean Sea. It occurs frequently in Sephardic and Ashkenazi Jews and their apparent immunity to the plague could account for the persecution that they suffered each time there was an epidemic of the disease.

A surprising clinical manifestation of both pneumonic and bubonic plague is the very rapid transition from an absence of symptoms to severe inflammation and fatal sepsis. During the pre-inflammatory phase, the bacteria migrate to the lymph nodes (bubonic plague) or the lungs (pneumonic plague) and multiply extensively. This early immune evasion is exerted by Yops, the absence of pathogen-associated molecular patterns, and an ability to survive in phagosomes.

We began this chapter with an account of biological warfare in the fourteenth century involving the causative organism of plague. More recently, during the Second World War, the Japanese dropped *Yersinia pestis*-infested fleas from low-flying planes on Chinese civilian populations resulting in limited outbreaks of bubonic plague. In the 1960s both the United States and the Soviet Union had programmes dedicated to weaponizing *Yersinia pestis*. What is particularly worrying is that this organism can be made resistant to multiple antibiotics, grown easily in large volumes, and can be stored almost indefinitely without losing its virulence. For this reason, it is a major concern to military and civilian health authorities.

Key points

- The bacterium *Yersinia pestis* causes bubonic plague (infection of lymph glands) when transmitted by fleas and pneumonic plague if transmitted by aerosols from a person with a lung infection.
- *Yersinia pestis* was responsible for three great plague pandemics: the Justinian plague (541–750), the Black Death (1347–1700s) and the Third

Pandemic (1894–mid-twentieth century). aDNA of *Yersinia pestis* has been recovered from teeth of victims of each of these plagues.

- *Yersinia pestis* diverged from the much less pathogenic *Yersinia pseudotuberculosis* about 5,700 years ago.

- Following divergence, *Yersinia pestis* acquired a number of plasmids that carry key virulence determinants. These included the *ymt* gene, essential for bacterial growth in the flea gut, the *pla* gene, necessary for the disease to spread in the body, and a type III secretion system.

- Key virulence aspects of *Yersinia pestis* are immune evasion mediated by outer membrane proteins, an absence of pathogen-associated molecular patterns, and an ability to survive in phagosomes.

Suggested Reading

Demeure C.E., Dussurget O., Mas Fiol G., Le Guern A-S., Savin C., and Pizarro-Cerdá J. (2019) *Yersinia pestis* and plague: an updated view on evolution, virulence determinants, immune subversion, vaccination, and diagnostics. *Genes & Immunity* 20, 357–70.

Dennis D.T. (2009) Plague as a biological weapon. doi:10.1007/978-1-4419-1266-4_2

Drancourt M., Aboudharam G., Signoli M., Dutour O., and Raoult D. (1998) Detection of 400-year-old *Yersinia pestis* DNA in human dental pulp: an approach to the diagnosis of ancient septicaemia. *Proceedings of the National Academy of Sciences* 95, 12637–40.

Frith J. (2012) The history of plague—Part1. The three great pandemics. *Journal of Military and Veterans Health* 20, 11–16.

Frith J. (2012) The history of plague—Part 2. The discoveries of the plague bacillus and its vector. *Journal of Military and Veterans Health* 20, 4–8.

Park Y.H., Remmers E.F., Lee W., Ombrello A.K., Chung L.K., et al. (2020) Ancient familial Mediterranean fever mutations in human pyrin and resistance to *Yersinia pestis*. *Nature Immunology* 21, 857–67.

Rasmussen S., Allentoft M.E., Nielsen K., Orlando L., Sikora M., et al. (2015) Early divergent strains of *Yersinia pestis* in Eurasia 5,000 years ago. *Cell* 163, 571–82.

Spyrou M.A., Tukhbatova R.I., Feldman M., Drath J., Kacki S., et al. (2016) Historical *Y. pestis* genomes reveal the European Black Death as the source of ancient and modern plague pandemics. *Cell Host & Microbe* 19, 874–81.

Valles X., Stenseth N.C., Demeure C., Horby P., Mead P.S., et al. (2020) Human plague: an old scourge that needs new answers. *PLoS Neglected Tropical Diseases* 14(8) doi: 10.1371/journal.pntd.0008251

4

A Multifaceted Pathogen: *Escherichia coli*

This is the first of three chapters considering members of the family Enterobacteriaceae: the other two cover *Salmonella enterica* and *Klebsiella pneumoniae*. Originally, members of this family were 'defined' by their ability to grow on MacConkey agar, a medium designed to selectively isolate Gram-negative and enteric (normally found in the intestinal tract) bacteria and differentiate them based on lactose fermentation. In the 5th edition of the *Manual of Clinical Microbiology* published in 1991 the Enterobacteriaceae contained many bacteria which differed greatly from *Escherichia coli* including species of *Pasteurella*, *Serratia*, *Proteus*, and *Providencia*. With the advent of genome sequencing these anomalous genera have been assigned to other families. However, as this and the two succeeding chapters will show, there is a remarkable diversity in the natural history of the extant members of the family Enterobacteriaceae.

Escherichia coli is a bacterium that is commonly found in the lower intestine of warm-blooded organisms. Because it can be grown and cultured easily and inexpensively in a laboratory setting, it usually is the first bacterium to which trainee microbiologists are introduced. It forms about 0.1% of the total bacterial load in the gut, where it benefits its hosts by producing vitamin K2 and preventing colonization of the intestine with pathogenic bacteria. It is expelled into the environment within faecal matter. For a long time, it was thought that *Escherichia coli* was a harmless commensal of the human gut but it now is recognized as a major cause of morbidity and mortality (Box 4.1). Based on specific virulence factors and pathogenicity processes, pathogenic *Escherichia coli* can be subdivided into different pathogroups that can be broadly grouped as intestinal and extraintestinal (Figure 4.1 and Table 4.1). Intestinal pathogens include at least seven major pathotypes differing in virulence mechanisms, infectious processes, and damage to target cells: enteropathogenic (EPEC), Shiga toxin-producing (STEC), enterotoxigenic (ETEC), enteroinvasive (EIEC), enteroaggregative (EAEC), diffusely adherent

Microbiology of Infectious Disease. Sandy B. Primrose, Oxford University Press.
© Sandy B. Primrose (2022). DOI: 10.1093/oso/9780192863843.003.0004

(DAEC), and adherent invasive (AIEC). The extraintestinal pathogens belong to two different pathotypes targeting different body compartments: uropathogenic (UPEC) and neonatal meningitis (NMEC). It now is generally recognized that bacteria formerly classified as species of *Shigella* are really human pathotypes of *Escherichia coli* and most closely resemble EIEC.

The genomes of more than 300 strains of *Escherichia* and *Shigella* have been completely sequenced and comparison of these sequences shows a remarkable amount of diversity. Only about 20% of each genome represents sequences present in every one of the isolates (core genome), while around 80% of each genome can vary among isolates. Each individual genome contains between 4,000 and 5,500 genes, but the total number of different genes among all of the sequenced strains (the pangenome) exceeds 16,000. This very large variety of component genes has been interpreted to mean that

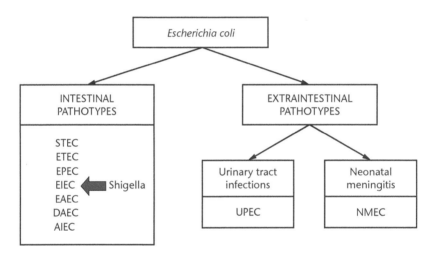

Figure 4.1. The different pathotypes of *Escherichia coli.*

Table 4.1. Summary of key virulence determinants in the commonest intestinal pathotypes.

Pathotype	Key virulence determinants
EPEC	Locus of enterocyte effacement carried on a pathogenicity island
ETEC	Produce heat-labile and/or heat-stable toxin plus colonization factors
STEC	Produce Shiga toxin (like heat-labile toxin of ETEC) as a result of lysogenic conversion
EIEC/*Shigella*	Has the pINV plasmid that carries a pathogenicity island encoding a T3SS and numerous effectors. Also has undergone pathoadaptation

two-thirds of the *Escherichia coli* pangenome originated in other species and arrived through the process of horizontal gene transfer.

Box 4.1 INTESTINAL PATHOTYPES OF *ESCHERICHIA COLI*

STEC produce Shiga toxin that causes inflammatory responses in target cells of the gut, leaving behind lesions which result in a characteristic bloody diarrhoea. This toxin can cause premature destruction of the red blood cells which then clog the kidneys, causing haemolytic-uremic syndrome (HUS).

ETEC are the most common cause of traveller's diarrhoea. The bacteria typically are transmitted through contaminated food or drinking water and adhere to the intestinal lining using various colonization factors (CFs) and then produce one of two proteinaceous enterotoxins. The larger of the two proteins, LT enterotoxin, is similar to cholera toxin in structure and function. The smaller protein, ST enterotoxin, causes cGMP accumulation in the target cells and a subsequent secretion of fluid and electrolytes into the intestinal lumen.

EPEC have an array of virulence factors that are similar to those found in *Shigella*. Adherence to the intestinal mucosa causes a rearrangement of actin in host cells resulting in significant deformation. EPEC cells are moderately invasive and elicit an inflammatory response. Changes in intestinal cell ultrastructure due to 'attachment and effacement' is likely to be the prime cause of diarrhoea.

EIEC and *Shigella* species cause a syndrome characterized by profuse diarrhoea, often bloody, and high fever. They are highly invasive and severely damage the intestinal wall through mechanical cell destruction.

EAEC have hair-like surface structures called fimbriae which can aggregate cells in tissue culture. They bind to the intestinal mucosa to cause watery diarrhoea through the production of an ST enterotoxin, similar to that of ETEC, and a haemolysin.

AIEC are able to invade intestinal epithelial cells and replicate intracellularly. They may be able to proliferate more effectively in hosts with defective innate immunity. They are associated with the ileal mucosa in Crohn's disease.

DAEC can cause diarrhoea in young children as well as urinary tract infections. They are characterised by the production of proteins known as the Afa/Dr adhesins.

UPEC are the cause of the vast majority of urinary tract infections (UTIs), including cystitis and pyelonephritis. UPEC express a multitude of virulence factors to breach the mucosal barrier triggering host inflammatory responses leading to cytokine production, neutrophil influx, and the exfoliation of infected bladder epithelial cells.

NMEC have the ability to survive in blood and invade meninges of infants to cause meningitis and are one of the most common infections that accounts for high mortality and morbidity rates (10–30%) during the neonatal period.

A hallmark phenotype of enteropathogenic *Escherichia coli* (EPEC) is the induction of a distinctive histopathology known as the attaching and effacing (A/E) lesion, which is characterized by the intimate adherence of bacteria to enterocytes, a signalling cascade leading to brush border and microvilli destruction, and loss of ions causing severe diarrhoea. EPEC contain a pathogenicity island called the 'locus of enterocyte effacement'. If this pathogenicity island is cloned into commensal strains of *Escherichia coli* then these acquire the EPEC phenotype showing that this pathogenicity island is a functional cassette both necessary and sufficient for EPEC virulence.

Enterotoxigenic *Escherichia coli* (ETEC) are defined by the ability to produce a heat-labile toxin (LT) and/or heat-stable toxin (ST) that stimulate fluid hypersecretion from intestinal cells thereby causing watery diarrhoea. ETEC also produce one or more cell adhesion factors, called colonization factors, that are essential for the initiation of disease. In addition to the genes for these factors, another 100 virulence genes have been identified and all are carried on a transmissible plasmid. Shiga toxin-producing *Escherichia coli* (STEC) are one of the commonest causes of food poisoning. They produce a toxin similar to that produced by *Shigella dysenteria* and the heat-labile toxin of ETEC. Unlike ETEC, the Shiga-like toxin is encoded by a transducing bacteriophage known as lambda (λ). Originally STEC were confined to *Escherichia coli* strain O157 but other strains have acquired pathogenicity following infection with bacteriophage λ (lysogenic conversion).

The major difference between commensal strains of *Escherichia coli* on the one hand and EIEC and *Shigella* species on the other is the presence of the plasmid pINV in the latter. Genes present on pINV enable the bacterium to invade intestinal epithelial cells, escape into the host cell cytosol, undergo cell-to-cell spread, and induce cell lysis in macrophages. Most of the virulence genes on pINV are located in a 30 kb pathogenicity island (PAI). This encodes components of a type III secretion system (T3SS), a molecular syringe that delivers bacterial effector proteins into the host cytoplasm, as well as most secreted effectors. Expression of the T3SS is highly regulated and responds to specific environmental cues such as temperature. These cues allow it to distinguish between free-living and host-associated environments. Synthesis of the T3SS is activated by a rise in temperature to 37 °C as would occur in the gastrointestinal tract. In addition, pINV carries other virulence genes and a key one is *icsA* that controls actin-mediated motility. The pathogenesis of diarrhoea induced by *Shigella* species involves bacterial invasion and spread through the colonic mucosa. The bacteria grow, divide, and move through the colonic epithelial cell. Moving bacteria are associated with polarized 'comet tails' rich in filaments of the host cell cytoskeletal protein actin.

The evolution toward pathogenic phenotypes by many bacteria, and *Escherichia coli* is no exception, is determined mainly by two mechanisms. The first of these is the acquisition of virulence genes by horizontal gene transfer of plasmids, such as pINV, and pathogenicity islands such as SHI-1, which carries several toxins, and SHI-2, which encodes bactericidal and immune evasion genes. The second mechanism, known as pathoadaptation, is the inactivation or loss of several chromosomal genes which negatively interfere with the expression of virulence factors required for survival within the host. While the first mechanism plays a crucial role in the colonization of a new host environment, the latter strongly contributes to the evolution of bacteria toward a more pathogenic phenotype.

Many pathoadaptive mutations are found in both EIEC and *Shigella*. These include the deletion of the *ompT* gene which encodes a protease that degrades the IcsA protein and therefore negatively interferes with host cell invasion by drastically reducing the ability of *Shigella* to spread into adjacent epithelial cells. Another typical pathoadaptive mutation of *Shigella* and EIEC is the inability to catabolise lysine due to the silencing of lysine decarboxylase activity. The lack of lysine decarboxylase prevents the synthesis of cadaverine, a polyamine that interferes with pathogenicity by blocking the release of bacteria into the cytoplasm of the infected cells. Another noteworthy pathoadaptive mutation in *Shigella* is the inactivation of the genes involved in the synthesis of nicotinic acid. This, in turn, prevents the conversion of aspartic acid to quinolinate, a compound that inhibits invasion and cell-to-cell spread of the bacteria.

Urinary tract infection (UTI) is the most commonly diagnosed urological and renal disease and is associated with morbidity in both hospitalized as well as outpatients (Box 4.2). In 50–90% of all uncomplicated UTIs, uropathogenic *Escherichia coli* (UPEC) is the most common organism seen. UPEC are strains of *Escherichia coli* that divert from their commensal status as intestinal flora and grow and persist in the urinary tract. To establish infection, the bacteria have to overcome several defence strategies of the host, including the urine flow, exfoliation of urothelial cells, endogenous antimicrobial factors, and invading neutrophils. Thus, UPEC harbour a number of virulence and fitness factors enabling the bacterium to resist and overcome these different defence mechanisms. Many of these virulence factors are found in EPEC, ETEC, EIEC, and *Shigella*. However, there is no single factor or combination of factors which allows the identification of UPEC among the commensal faecal flora apart from the ability to enter the urinary tract and cause an infection. Consequently, UTI pathogenesis is assumed to be determined by a complex interplay of bacterial and host factors.

Box 4.2 URINARY TRACT INFECTIONS (UTIS)

A UTI is an infection that affects part of the urinary tract. When it affects the lower urinary tract it is known as a bladder infection (cystitis) and when it affects the upper urinary tract it is known as a kidney infection (pyelonephritis). Symptoms of cystitis include pain during urination (dysuria), frequent urination (frequency), and feeling the need to urinate despite having an empty bladder. People experiencing pyelonephritis, which is a much more serious condition, may experience flank pain, fever, and nausea and vomiting in addition to the classic symptoms of a lower urinary tract infection. A predisposition for bladder infections may run in families. Other risk factors include diabetes and having an enlarged prostate.

The bacteria that cause UTIs typically enter the bladder via the urethra, having been transmitted there from the bowel. Women are more prone to UTIs than men because, in females, the urethra is much shorter and closer to the anus. As a woman's oestrogen levels decrease with menopause, her risk of urinary tract infections increases due to the loss of protective vaginal flora. Chronic prostatitis may cause recurrent urinary tract infections in males and the risk of infections increases as males age. Urinary catheterization increases the risk for UTIs and the risk of bacteriuria (bacteria in the urine) is between 3% and 6% per day.

Bacteria colonizing the urinary tract have to overcome the shear stress of urine flow. This stress can vary considerably. In the bladder, the flow changes dramatically upon voiding whereas in the renal tubules more subtle variations occur as the body regulates renal function. For successful colonization in this hydrodynamically challenging environment, attachment of the UPEC to the epithelium is essential. Major roles in attachment have been ascribed to two hair-like appendages on the bacteria surface: type 1 and P fimbriae. The latter mediate binding between the bacteria and the epithelial cells lining the tubules, while type 1 appear to play a role in inter-bacterial binding and biofilm formation. If a biofilm forms it may cause blockage of the renal tubules and severe kidney problems. Antibiotics are prescribed for the treatment of pyelonephritis and sometimes for cystitis but there is an alternative treatment. Type I fimbriae attach to mannose residues on the surface of urinary epithelial cells. When taken orally, D-mannose is excreted unchanged by the kidneys and binds to the type I fimbriae preventing epithelial attachment by the UPEC.

Escherichia coli is a leading cause of neonatal bacterial meningitis (NMEC) but the mechanism of pathogenesis remains elusive, as it does for UPEC. Penetration of the blood–brain barrier is a critical step for development of meningitis and requires a functioning bacterial *ibeA* gene. Numerous other virulence-associated genes have been found in NMEC, many on plasmids, that do not occur in UPEC or commensal strains of *Escherichia coli* but their function is unknown.

As should be clear from the foregoing, *Escherichia coli* truly is a multifaceted pathogen with its ability to cause different forms of intestinal disorder by various mechanisms as well as UTIs and neonatal meningitis. A common feature of all the pathovars is a host of virulence factors, some unique to a particular pathovar and others shared between pathovars. The function of these virulence genes, which mostly are carried on plasmids, largely remains obscure. Although *Shigella* is considered a pathovar of *Escherichia coli*, clinical microbiologists maintain the distinction on the basis of the severity of symptoms caused by the former. This is not unreasonable for *Shigella*, and many EIEC have another strategy for increasing virulence: pathoadaptation by inactivation or loss of chromosomal genes whose expression reduces virulence. It should come as no surprise that *Shigella* species harbour large numbers of insertion sequences, mobile pieces of DNA that can insert at random through the genome. Given the diversity of pathogenicity determinants one can ask what it would take to turn a commensal strain of *Escherichia coli* into a pathogen. At its simplest, this would involve acquisition of the EPEC pathogenicity island known as the locus of enterocyte effacement or perhaps the bacteriophage carrying the Shiga-like toxin.

Key points

- *Escherichia coli* normally behaves as a commensal organism and constitutes about 0.1% of the normal flora of the human gut where its beneficial properties include synthesis of vitamin K.

- Some strains of *Escherichia coli* can cause disease and these pathogenic strains have acquired various pathogenicity determinants, most likely by horizontal gene transfer.

- The diseases caused by *Escherichia coli* fall into two types: intestinal diseases and non-intestinal diseases. The latter type includes urinary tract infections and neonatal meningitis.

- There are seven different types of intestinal disease caused by *Escherichia coli* and each type is caused by a different pathovar with unique virulence determinants.

- Based on genome analysis, the bacteria formerly classified as *Shigella* species are simply pathovars of *Escherichia coli*.

- Some *Escherichia coli* strains can increase their virulence by inactivating certain chromosomal genes, a process known as pathoadaptation, and this may occur by movement of insertion sequences.

Suggested Reading

Belotserkovsky I. and Sansonetti P.J. (2018) Shigella and enteroinvasive *Escherichia coli*. *Current Topics in Microbiology and Immunology* **416**, 1–26.

Croxen M.A. and Finlay B.B. (2010) Molecular mechanisms of *Escherichia coli* pathogenicity. *Nature Reviews Microbiology* **8**, 26–38.

Kaper J.B., Nataro J.P. & Mobley H.T. (2004) Pathogenic *Escherichia coli. Nature Reviews Microbiology* **2**, 123–40.

Pasqua M., Michelacci V., Di Martino M.L., Tozzoli R., Grossi M., et al. (2017) The intriguing evolutionary journey of enteroinvasive *E. coli* (EIEC) toward pathogenicity. *Frontiers in Microbiology* **8**, 2390. doi:10.3389/fmicb. 2017.02390

Sarowska J., Futoma-Koloch B., Jama-Kmiecik A., Frej-Madrzak M., Ksiazczyk M., et al. (2019 Virulence factors, prevalence and potential transmission of extraintestinal pathogenic *Escherichia coli* isolated from different sources: recent reports.) *Gut Pathogens* **11**, 10. doi:10.1186/s13099-019-0290-0

5

Fever and Food Poisoning: The Two Faces of *Salmonella*

The taxonomy and nomenclature of *Salmonella* is very confusing: much more so than for any other genus of bacteria. Traditionally, *Salmonella* isolates were named according to clinical considerations. For example, *Salmonella typhi* caused typhoid fever and *Salmonella paratyphi* caused paratyphoid fever. When it was discovered that strains with different biochemical properties (phenotypes) caused similar types of disease, new strains received names according to the place where they were first isolated, for example *Salmonella Dublin, Salmonella Montevideo*. Overlying this was the division of *Salmonella* into serotypes according to the Kauffman–White system that is based on surface antigens. First the 'O' antigen type is determined based on oligosaccharides associated with lipopolysaccharide in the cell wall and then the 'H' antigen is determined based on flagellar proteins.

With the advent of classification based on DNA sequence analysis, the genus *Salmonella* now is divided into just two species: *Salmonella bongori*, which infects cold-blooded animals, and *Salmonella enterica*, which infects humans and other warm-blooded animals. *Salmonella enterica* is divided into seven subspecies of which *Salmonella enterica* subs. *enterica* is the only one considered here because it is the only one that can infect mammals. The subspecies then are divided into ~ 2,500 serotypes such as *Salmonella enterica* Typhimurium and *Salmonella enterica* Typhi (Figure 5.1). Although this nomenclature has been formally adopted, many older microbiologists still use traditional names which also will be found in many less recent publications.

Infections with *Salmonella* are of two types: non-typhoidal and typhoidal. Non-typhoidal serotypes such as *Salmonella* Typhimurium and *Salmonella* Enteriditis cause food poisoning. They can infect a range of animals and are zoonotic. The number of cases worldwide of food poisoning caused by non-typhoidal serotypes is not known but is estimated to be over 100 million annually. The sources of infection for *Salmonella* food poisoning are twofold:

Microbiology of Infectious Disease. Sandy B. Primrose, Oxford University Press.
© Sandy B. Primrose (2022). DOI: 10.1093/oso/9780192863843.003.0005

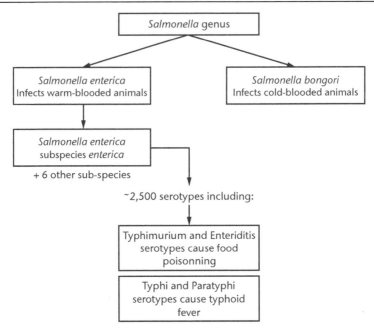

Figure 5.1. The taxonomic structure of the genus *Salmonella*.

poor hygiene of food handlers and zoonotic through consumption of contaminated food, particularly chicken, hen's eggs, and pork. In recent years, zoonotic transmission from poultry has been greatly reduced through adoption by the food industry of better procedures and vaccination of hens.

Typhoid fever is caused by serotypes which only infect humans and the principal ones are *Salmonella* Typhi and *Salmonella* Paratyphi A, B, or C. The incidence of typhoid fever in the Western world is very low and affected individuals have usually acquired it from under-developed countries although there have been some notable exceptions (Box 5.1). Globally, typhoid fever is a serious infectious disease, particularly in places with poor sanitary conditions, and annually there are about 26 million cases and over 200,000 deaths. Because there are no zoonotic reservoirs of the typhoidal serotypes, the source of most infections is the consumption of contaminated water. This means that the bacteria must be able to adapt to life in water, an environment that will be nutrient poor and at a lower temperature than the human body. In the absence of an animal model for *Salmonella* Typhi infection, bacteria were transferred from laboratory culture media to water and the changes in gene expression monitored. The key observations were that the expression of genes involved in catabolism and ATP generation increased whilst those involved in biosynthesis decreased. Put simply, the bacteria went into survival mode.

Box 5.1 NOTABLE TYPHOID OUTBREAKS

In March 1963, an outbreak of typhoid occurred in the Swiss ski resort of Zermatt. It coincided with a prestigious ski competition that attracted skiers and spectators from around the world. By the time the epidemic was over, 437 people had contracted the disease and 3 had died. An investigation revealed that the sanitation systems in the town were not fit for purpose and, subsequently, state-of-the-art drinking water and wastewater systems were installed.

In 1964 there was an outbreak of typhoid fever in the Scottish town of Aberdeen and over 400 people were infected. This outbreak was traced to a catering-size tin of corned beef that had been manufactured in Uruguay. After cooking, the cans were quickly cooled with raw water from the Uruguay River. It is believed that contaminated water was sucked into one of the cans through a defective seal. Once the infected can arrived at a grocer's shop it was opened and, not looking tainted in any way, was sliced on a machine. Most of the victims had eaten either the corned beef or other meat sliced on the same machine.

In 1973, 188 workers at a migrant labour camp in Dade County, Florida contracted typhoid from contaminated water. A more famous US case was that of Mary Mallon (1869–1938), better known as 'Typhoid Mary', who was the first asymptomatic carrier of typhoid to be discovered. She was a cook and infected at least 122 people, of whom 5 died, indicating a lack of good hygienic practices.

Human infections with both typhoidal and non-typhoidal serotypes occur following ingestion. Both must cross the barrier created by the intestinal cell wall but they use different strategies to cause infection. Non-typhoidal serotypes infect M cells which are specialized epithelial cells of intestinal lymphoid tissues such as Peyer's patches. Infection is by bacterial-mediated endocytosis and this gives rise to inflammation. The bacteria also disrupt the tight junctions between intestinal cells leading to excessive flow of ions and water into and out of the intestine, that is, diarrhoea. Whereas the non-typhoidal serotypes cause localized infection, the typhoidal serotypes cause a systemic infection. Typhoidal serotypes pass through the lymphatic system of the intestine and into the blood of patients to cause fever. They also are transported to various organs where they may cause sepsis.

Salmonella is believed to have diverged from a common ancestor with *Escherichia coli* (see p31) about 100 million years ago. Nevertheless, genome sequencing has shown that they have a similar chromosomal gene structure. The genomes of *Salmonella* Typhimurium and *Salmonella* Typhi are very similar in size (4.8 Mbp) and they share about 90% of their genes and, consequently, many pathogenic traits. For example, both serovars encode two type III protein secretion systems (T3SSs) within *Salmonella* pathogenicity islands 1 (SPI-1) and 2 (SPI-2), which mediate their close interactions with host cells. Through the activity of the several effector proteins they deliver, these T3SSs mediate bacterial entry, intracellular replication, and the transcriptional reprogramming of the target cells by subverting the cellular machineries that control actin cytoskeleton dynamics and signal transduction. Despite the highly conserved nature of these T3SSs, the composite of effector proteins that they deliver differs significantly between typhoidal and non-typhoidal serotypes. One key difference is the absence of two effector proteins, GtgE and SopD2, from typhoidal serotypes. These proteins target Rab32, a component of cells that restrict intracellular pathogen replication, and their absence prevents *Salmonella* Typhi replication in non-human hosts.

Unique factors to *Salmonella* Typhi include typhoid toxin and the Vi capsular polysaccharide. The typhoid toxin belongs to the AB_5 family of toxins (see also p57) consisting of enzymatic A subunits and receptor-binding B subunits. One of the A subunits is the product of the *CdtB* gene and encodes a deoxyribonuclease which causes DNA damage and cell cycle arrest. A key feature of the toxin is that it is secreted from infected cells and can be distributed systemically. The importance of the typhoid toxin is illustrated by the fact that it reproduces many of the symptoms of typhoid fever when administered to experimental animals. The Vi capsular polysaccharide, which is encoded by a group of genes on a *Salmonella* pathogenicity island (SPI-7) is thought to allow the pathogen to escape the host's immune surveillance. Furthermore, through the process of host adaptation, *Salmonella* Typhi has lost a number of genes that are present in non-typhoidal *Salmonella* serovars. For example, *Salmonella* Typhi has lost the ability to synthesize the amino acids tryptophan and cysteine. Also, many of the genes involved in intestinal colonization by *Salmonella* Typhimurium are lost in *Salmonella* Typhi.

Between 3% and 5% of patients who recover from typhoid become chronic, asymptomatic carriers of *Salmonella* Typhi and they can infect others (like 'Typhoid Mary', Box 5.1) if they do not practise good hygiene. The bacterial reservoir in carriers is the gall bladder. Studies have shown a strong correlation between the presence of gallstones and the chance of progressing to a chronic carrier state and this is because the bacteria can form a biofilm on the gallstones. A number of genes are induced in *Salmonella* Typhi, but not *Salmonella* Typhimurium, in the presence of bile and these include the ones involved in the synthesis of the Curli protein. This protein is similar to the amyloid fibres seen in higher eukaryotes and is involved in cell aggregation and cell adhesion.

The genomes of a large number of isolates of *Salmonella* Typhi from different parts of the world have been sequenced. Remarkably, they contain relatively little variation throughout the entire genome with the exception of antibiotic resistance and particularly resistance to fluoroquinolones (see p57). This lack of diversity suggests that all extant strains arose from a single predecessor, possibly 50,000 years ago, and that only a few, very closely related strains are responsible for all cases of typhoid fever. Mutation analysis has shown that most of them belong to what is known as the H58 haplotype which is characterized by a point mutation in the gene for glycerol-3-phosphate dehydrogenase. This mutation may have nothing to do with the global dominance of H58 and its presence could be purely coincidental but it was not present in older isolates.

In 1948 it was discovered that all typhoid strains were susceptible to the antibiotic chloramphenicol and the incidence of severe complications dropped dramatically. It was twenty-five years before chloramphenicol resistance became established in *Salmonella* Typhi, in marked contrast to non-typhoidal strains of *Salmonella* where antibiotic resistance plasmids were commonplace. Following the development of chloramphenicol resistance, amoxicillin, ampicillin, and trimethoprim-sulfamethoxazole treatment became standard treatments and when resistance to them developed they were replaced with fluoroquinolones such as ciprofloxacin. Fluoroquinolone resistance usually is due to mutations in *gyrA*, the gene for DNA gyrase which is the antibiotic target (Box 5.2). Unlike all other genes in *Salmonella* Typhi,

---

Box 5.2 QUINOLONE ANTIBIOTICS

Nearly all quinolone antibiotics in use are fluoroquinolones, which contain a fluorine atom in their chemical structure (Figure 5.2), and these are effective against both Gram-negative and Gram-positive bacteria. Quinolones all have a name that ends in *-oxacin* and the best known and most widely used is ciprofloxacin (Figure 5.2). They interfere with DNA replication by preventing bacterial DNA from unwinding and duplicating. Specifically, they inhibit the ligase activity of the type II topoisomerases, DNA gyrase and topoisomerase IV, which cut DNA to introduce supercoiling, while leaving nuclease activity unaffected. With the ligase activity disrupted, these enzymes release DNA with single- and double-strand breaks resulting in cell death.

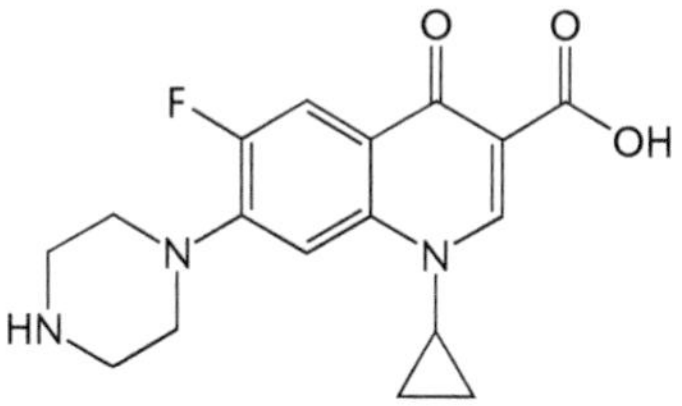

Figure 5.2. Schematic representation of a fluoroquinolone antibiotic.

Three mechanisms of resistance are known. Some bacteria have efflux pumps that can act to decrease intracellular quinolone concentration. In Gram-negative bacteria, plasmid-mediated resistance genes produce proteins that can bind to DNA gyrase, protecting it from the action of quinolones. This type of plasmid-borne resistance has been seen in recent isolates of *Salmonella* Typhi. Finally, mutations at key sites in DNA gyrase or topoisomerase IV can decrease their binding affinity to quinolones, decreasing the drugs' effectiveness. Such changes in DNA gyrase are commonplace in isolates of *Salmonella* Typhi.

at least fifteen different mutations in *gyrA* have been found in isolates from different parts of the world. Worryingly, strains that also are resistant to third-generation cephalosporins have been isolated and the treatment options for these extremely resistant (XDR) strains are rapidly disappearing.

As noted earlier, *Salmonella* Typhi globally has developed resistance to many antibiotics and in the absence of effective therapy the only option is prevention of the disease through vaccination. Several different vaccines are available. Two of them (Typhim VI and Typherex) are based on purified Vi capsular polysaccharide, which is a known virulence factor, and also give protection against *Salmonella* Paratyphi C. Improved vaccines have been developed where the immune response is enhanced by coupling the Vi polysaccharide to tetanus toxoid or a non-toxic form of *Pseudomonas aeruginosa* exotoxin A. Vivotif is a live strain of *Salmonella* Typhi strain Ty21a that also offers some degree of protection against *Salmonella* Paratyphi A and B. It was attenuated by mutation of multiple genes including ones for the production of the Vi capsular polysaccharide. The biggest problem with these vaccines is that they do not give long-lasting immunity and have to be read-ministered every few years. This is not a problem for travellers to areas where typhoid is endemic but is a major barrier to disease prevention in developing countries.

In contrast to typhoidal strains, non-typhoidal serotypes of *Salmonella* show much greater genomic variation. There are many mutational differences between strains isolated from different parts of the world and considerable variation in the clinical symptoms observed in infections. In developed countries, both *Salmonella* Typhimurium and *Salmonella* Enteritidis cause self-limiting enterocolitis (food poisoning). However, in the past two decades, both serotypes have emerged as a cause of a life-threatening invasive bacterial disease in many countries in sub-Saharan Africa. This disease occurs in individuals with malnutrition, severe malaria, and advanced HIV infection/AIDS. Genomic analysis of the strains causing invasive non-typhoidal *Salmonella* disease has shown that there were just a few clonal lineages and they have undergone genomic degradation and lost many of the genes normally found in *Salmonella* Typhimurium but not *Salmonella* Typhi. That is, they have undergone pathoadaptation.

More than 80% of outbreaks of *Salmonella* food poisoning in Europe have been traced to the consumption of inadequately cooked eggs or foods containing uncooked eggs. In almost all of the cases *Salmonella* Enteriditis is the culprit and it now is common practice to vaccinate chicken flocks with attenuated strains. Often there are outbreaks of *Salmonella* food poisoning that are not linked to eggs and the challenge is to identify the source of the outbreak. It is known that particular non-typhoidal strains are associated with particular foodstuffs but which strain is causing a particular outbreak? Genome sequencing now is so rapid that the causative strain can be identified within twenty-four hours and this gives a strong indication of the type of food that is the source of infection.

Key points

- There are two species of *Salmonella* and only *Salmonella enterica* infects warm-blooded animals. The sub-species *Salmonella enterica* subsp. *enterica* is the only one to infect humans and other mammals.
- Infections of humans are the result of the ingestion of certain serotypes of *Salmonella enterica* subsp. *enterica*. Typhoidal serotypes cause typhoid fever and non-typhoidal serotypes cause food poisoning.
- Both typhoidal and non-typhoidal serotypes need to cross the wall of the intestine but do so in different ways and using different effectors.
- Typhoidal serotypes produce typhoid toxin and pathogenicity is enhanced by the presence of the Vi antigen. They also have undergone genomic degradation by losing the function of certain biosynthetic genes present in non-typhoidal serotypes.
- There is little genomic variation in typhoidal serotypes isolated from different parts of the world whereas non-typhoidal serotypes show a great deal of variation.
- In sub-Saharan Africa there is evidence that non-typhoidal serotypes are undergoing genomic degradation and becoming invasive.

Suggested Reading

Feasey N.A., Hadfield J., Keddy K.H., Dallman T.J., Jacobs J., et al. (2016) Distinct *Salmonella* Enteritidis lineages associated with enterocolitis in high-income settings and invasive disease in low-income settings. *Nature Genetics* **48**, 1211–17.

Gal-Mor O., Boyle E.C., and Grassl G.A. (2014) Same species, different diseases: how and why typhoidal and non-typhoidal *Salmonella enterica* serovars differ. *Frontiers in Microbiology* 5, 391. doi:10.3389/fmicb.2014.00391

Harrell J.E., Hahn M.M., D'Souza S.J., Vasicek E.M., Sandala J.L., et al. (2021) *Salmonella* biofilm formation, chronic infection, and immunity within the intestine and hepatobiliary tract. *Frontiers in Cellular and Infection Microbiology* **10**, 624622. doi:10.3389/fcimb.2020.624622

Levine M.M. and Simon R. (2018) The gathering storm: is untreatable typhoid fever on the way? *mBio* MarApr, 9(2). doi:10.1128/mBio.00482-18

Sabbagh S.C., Forest C.G., Lepage C., Leclerc J-M., and Daigle F. (2010) So similar, yet so different: uncovering distinctive features in the genomes of *Salmonella enterica* serovars Typhimurium and Typhi. *FEMS Microbiology Letters* **305**, 1–13.

Tang Y., Davies R., and Petrovska L. (2019) Identification of genetic features for attenuation of two *Salmonella* Enteriditis vaccine strains and differentiation of these from wildtype isolates using whole genome sequencing. *Frontiers in Veterinary Science* **6**, 447. doi:10.3389/fvets.2019.00447

Wong V.K., Baker S., Pickard D.J., Parkhill J., Page A.J., et al. (2015) Phylogeographical analysis of the dominant multidrug-resistant H58 clade of *Salmonella* Typhi identifies inter- and intracontinental transmission events. *Nature Genetics* **47**, 632–9.

6

Understanding Pathogen Populations:
Klebsiella pneumoniae

Klebsiella pneumoniae is an opportunistic pathogen that is a major cause of nosocomial (hospital-acquired) infections. It is an opportunistic pathogen in that those most at risk tend to be vulnerable patient groups such as immunocompromised individuals, neonates, and the elderly, especially those with indwelling devices. The most common manifestations are pneumonia, urinary tract infections, and wound infections, any of which can progress to bacteraemia (infection of the bloodstream). There is a long history of these hospital-acquired infections (HAI) dating back over 100 years and for this reason are often referred to as 'classical' *Klebsiella pneumoniae* infections. Since the 1980s it has been recognized that *Klebsiella pneumoniae* can be a 'true' pathogen in that it causes infections in otherwise healthy individuals. These community-acquired infections (CAI) include infection of the eye (endophthalmitis), meningitis, necrotizing fasciitis (death of skin and soft tissues), and abscesses, all of which can progress to bacteraemia. There are similarities between *Salmonella enterica* subsp. *enterica*, with its typhoidal and non-typhoidal serotypes, and *Klebsiella pneumoniae* causing HAI and CAI. However, whereas *Salmonella* species are obligate pathogens, *Klebsiella pneumoniae* also can be found in a wide range of environmental niches (water, soil, plant surfaces) where it is known to exhibit extensive phenotypic and genetic diversity. This raises the question as to the genetic composition of different populations, especially the HAI and CAI populations.

Strains of *Klebsiella pneumoniae* that cause infections possess many virulence factors and the best characterized are the capsule, lipopolysaccharide, pili, and siderophores. The capsule is a polysaccharide matrix that coats the cell and its importance as a virulence factor is demonstrated by the near absence of pathogenicity when mutants lacking a capsule are tested in mouse models. The capsular polysaccharide is antigenic and is known as the K antigen. Serotypes K1 to K78 are found in HAI strains but only K1 and K2 are found in CAI strains. The capsule plays a key role in defending bacteria from

Microbiology of Infectious Disease. Sandy B. Primrose, Oxford University Press.
© Sandy B. Primrose (2022). DOI: 10.1093/oso/9780192863843.003.0006

host defences and does this in a number of ways. First, it prevents phagocytosis of the bacteria by immune cells. Second, it hinders the bactericidal action of various host-derived antimicrobial peptides such as defensins and lactoferrin by sequestering these molecules and stopping them reaching the cells. Third, it blocks components of the complement system from interacting with the membrane thereby preventing complement-mediated cell lysis.

The lipopolysaccharide that forms the outer membrane of Gram-negative bacteria prevents complement-mediated killing but the lipid component (lipid A) acts as a pathogen-associated molecular pattern (PAMP) that is recognized by the pattern recognition receptor TLR4. Stimulation of TLR4 leads to recruitment of neutrophils and macrophages that clear bacterial infections. Many strains of *Klebsiella pneumoniae* can modify their lipid A such that it induces a much weaker inflammatory response than bacteria with normal lipid A.

Most *Klebsiella pneumoniae* isolates express two types of adhesins, type 1 and type 3 pili (sometimes called fimbriae), that enable them to stick to epithelial cells and to form biofilms. Type 1 pili are found in the majority of members of the family Enterobacteriaceae such as *Escherichia coli*, *Shigella*, and *Salmonella* species. These pili mediate adhesion to mannose-containing structures and their expression is phase variable, that is expression is turned on or off depending on the orientation of an invertible DNA element known as *fim*. Type 3 pili are present in practically all *Klebsiella pneumoniae* isolates and, *in vitro* at least, mediate adhesion to several different cell types. The receptor for type 3 pili has not yet been identified. Historically, type 3 pili have not been associated with *Escherichia coli* but recently a number of strains have been isolated where type 3 pili synthesis is encoded on plasmids. In terms of pathogenicity, type 1 pili play a key role in urinary tract infections facilitating adhesion to the bladder epithelium. They play no other role in attachment to host cells. Type 3 pili do not appear to be involved in host colonization at all. Indeed, the presence of these pili facilitates binding of the bacteria to phagocytes and their destruction. However, where both pili play a role is in the formation of biofilms on indwelling medical devices such as catheters, endotracheal tubes, etc. These biofilms are major sources of HAI which probably result from the release of clumps of biofilm into the bladder, lungs, and other tissues.

As noted earlier (p20), iron is a key element for the growth of pathogens and a key host defence is to sequester iron in molecules such as transferrin. Pathogens respond to this need for iron by producing siderophores that can steal iron from transferrin or scavenge it from the environment. The principal siderophore in members of the family Enterobacteriaceae, including *Klebsiella pneumoniae*, is enterochelin which has a very high affinity for iron. However, the host can produce lipocalin-2 which can neutralize the iron-chelating ability of enterochelin. Where *Klebsiella pneumoniae* differs from its enteric relatives (*Escherichia coli, Salmonella*) is in its ability to produce multiple siderophores. One of these is salmochelin, a glycosylated form of enterochelin that is not neutralized by lipocalin-2 because it cannot be bound by it. Another siderophore found in *Klebsiella pneumoniae* is yersiniabactin that was first discovered in *Yersinia* species (p26) where it is encoded by a pathogenicity island. Yersiniabactin is not inhibited by lipocalin-2 and its presence enables *Klebsiella pneumoniae* to grow to high numbers in lungs. Finally, many CAI strains also produce aerobactins.

Strains of *Klebsiella pneumoniae* involved in CAIs are said to be hypervirulent because of the severity of the disease that they cause. Hypervirulence is associated with a number of particular virulence factors. First, the vast majority of hypervirulent strains have the K1 capsular serotype and the remainder have the K2 serotype. By contrast, strains causing HAIs can have any of the capsular serotypes although twenty-five out of the seventy-eight serotypes account for 70% of all clinical isolates. A key feature of cells with the K1 and K2 serotypes is that they are more resistant to phagocytosis and intracellular killing. Furthermore, many of the CAI strains produce a hypercapsule that is much larger and more viscous than ordinary capsules and is believed to contribute significantly to pathogenicity. Another feature of hypervirulent strains is that all of them produce more than one of the siderophores discussed earlier and in most cases produce three or all four of them. By contrast, only a minority of HAI strains produces more than one siderophore and if they do, it will most likely be yersiniabactin in addition to enterochelin. The presence of aerobactin is always associated with a hypercapsule because the genes for aerobactin synthesis are carried on the same virulence plasmid that encodes an enhancer of capsule production. Many hypervirulent strains also synthesize colibactin, a genotoxic compound that cross-links DNA and causes double-stranded DNA breaks. In experiments in mice, it has been shown that loss of colibactin synthesis reduces the spread of *Klebsiella pneumoniae* to the blood, liver, spleen, and brain.

All clinical isolates of *Klebsiella pneumoniae* are resistant to ampicillin, the β-lactam antibiotic (Box 6.1) of choice for treating infections caused by Gram-negative bacteria. This resistance stems from the presence of a chromosomal *bla* gene that encodes a β-lactamase that cleaves the β-lactam ring. The *bla* gene has been captured from the chromosome multiple times by a transposon and frequently can be found on plasmids. These mobile variants of the *bla* gene often carry mutations that results in an extended spectrum of activity that includes third-generation cephalosporins and even carbapenems. Some clinical isolates carry two copies of *bla*, a chromosomal version and a plasmid-borne version, the latter being more strongly expressed because it is under the control of a transposon promoter.

Box 6.1 β-LACTAM ANTIBIOTICS AND β-LACTAMASES

β-lactam antibiotics are antibiotics that contain a β-lactam ring in their molecular structure (Figure 6.1). This includes penicillin derivatives which were the first of this class, cephalosporins, monobactams, and carbapenems. They kill bacteria by inhibiting the synthesis of the peptidoglycan layer of bacterial cell walls. The final step in the synthesis of the peptidoglycan is facilitated by DD-transpeptidases, also known as penicillin-binding proteins (PBPs). The normal substrate for these enzymes is D-alanyl-D-alanine but β-lactams have a structural similarity, and their β-lactam nucleus can bind irreversibly to the active site of the enzyme and prevent normal peptidoglycan synthesis. β-lactamases are enzymes that confer resistance to the antibiotics by cleaving the β-lactam ring. As new β-lactam antibiotics have been introduced to counter resistance, and there have been six generations of cephalosporins developed, so the β-lactamase evolves. We now have extended spectrum β-lactamases (ESBL) that can destroy the activity of many of the current generation of cephalosporins. Carbapenems are resistant to ESBLs but are inactivated by other β-lactamases that *Klebsiella pneumoniae* has acquired by horizontal gene transfer. Unlike most β-lactamases that have serine at their active site, carbapenemases utilize zinc ions in the cleavage process. The most notorious of the carbapenemases is the New Dehli metallo-β-lactamase (NDM-1) that was isolated from patients in India in 2009 and now has spread around the world.

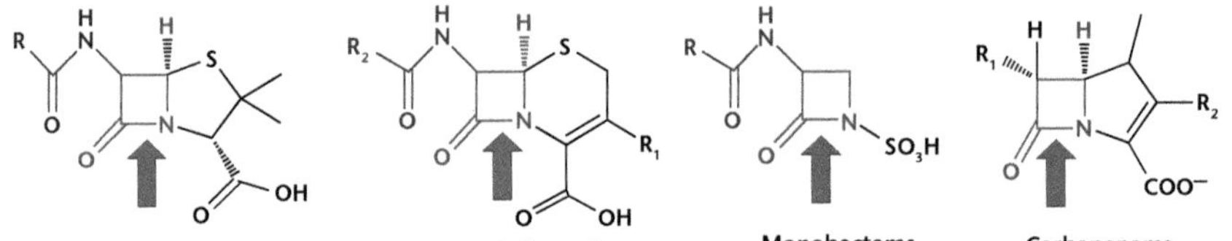

Figure 6.1. Representative examples of β-lactams. The arrows show the bond broken by β-lactamases.

β-lactams apart, CAI isolates of *Klebsiella pneumoniae* generally are sensitive to the commonly used antibiotics but the situation with HAI isolates is quite different. Many of them are resistant to aminoglycosides, fluoroquinolones, sulphonamides, tetracyclines, and trimethoprim and this resistance is the result of the acquisition of plasmids and transferable genetic elements. To date, over 400 antimicrobial resistance genes have been identified in different *Klebsiella pneumoniae* genomes, making the treatment of HAI very challenging. In summary, HAI isolates are opportunistic pathogens that are multi-drug resistant whereas CAI isolates are hypervirulent but treatable with antibiotics.

Klebsiella pneumoniae strains can be divided into different sequence types (STs) by multi-locus sequence typing using seven housekeeping genes and globally there are hundreds of different STs. A more sophisticated typing scheme is based on nucleotide sequence variation in hundreds of genes to identify clonal groups (CGs) and these mostly equate to STs. Among the isolates that cause HAI, particular STs or CGs cause localized problems within a single hospital or healthcare group but cause no or limited problems elsewhere. However, a small sub-set of lineages have spread globally, and one example is ST258. This strain arose by recombination between an ST11 strain and an ST442 strain and its genomic structure has been analysed in great detail, but this has provided no explanation for its global spread. By contrast, community-acquired infections result from the same set of sequence types regardless of their country of origin.

There are a number of key unanswered questions regarding *Klebsiella pneumoniae* infections. For example, why are community-acquired infections largely restricted to south-east Asia and Taiwan? The predisposing health conditions (diabetes, malignancies, alcoholism, and chronic pulmonary obstructive disease) leading to these infections are found in populations all over the world. Also, why is multi-drug resistance not a problem with these infections? What is the source of the strains causing healthcare-acquired infections? It has been suggested that these derive from bacteria living commensally in the gut or on the skin. However, if this were the case then a greater diversity of STs would be expected in any locality—unless only certain STs can cause infection. Finally, what is different about certain STs causing HAI that results in them spreading globally?

Key points

- *Klebsiella pneumoniae* is widespread in the environment and is a human commensal but only a small number of strains cause human infections.

- Human infections are of two types: hospital-acquired infections (HAI) and community-acquired infections (CAI).

- *Klebsiella pneumoniae* is resistant to ampicillin because it has a *bla* gene, but many clinical isolates are resistant to most β-lactam antibiotics because of mutations in the *bla* gene or by the acquisition of genes encoding other β-lactamases.

- Strains causing HAI are opportunistic pathogens but are difficult to treat because they are multi-drug resistant. Most of these strains circulate locally but a small number for unknown reasons have spread globally.

- Strains causing CAI are true pathogens and are hypervirulent but sensitive to most antibiotics. They are largely restricted to south-east Asia and Taiwan but the reason for this is not known.

Suggested Reading

Martin R.M. and Bachman M.A. (2018) Colonization, infection, and the accessory genome of *Klebsiella pneumoniae*. *Frontiers in Cellular and Infection Microbiology* **8**, 4. doi:10.3389/fcimb.2018.00004

Paczosa M.K. and Mecsas J. (2016) *Klebsiella pneumoniae*: going on the offense with a strong defense. *Microbiology and Molecular Biology Reviews* **80**, 629–61.

Wyres K.L., Lam M.M.C., and Holt K.E. (2020) Population genomics of *Klebsiella pneumoniae*. *Nature Reviews Microbiology* **9**, 2703. doi:10.1038/s41579-019-0315-1

7

A Surprising Pathogen: *Vibrio cholerae*

At first sight, cholera is a very simple disease. The bacterium *Vibrio cholerae* produces a toxin that causes chloride secretion from cells and acute secretory diarrhoea (Figure 7.1) resulting in the so-called rice water stools. Continued diarrhoeal purging in the absence of aggressive fluid replacement results in blood electrolyte imbalance, low blood pressure, and death. The story, as we shall see, is much more complex.

The symptoms of cholera are very distinctive and until the early nineteenth century, the disease was largely confined to the Indian sub-continent. In 1817, the disease started spreading worldwide and there have been seven global pandemics between then and today. During the third pandemic (1846–1860), and before the germ theory of disease had been developed, English physician John Snow was convinced that cholera was transmitted by contaminated water. He famously persuaded the local authorities in Soho in London to disable the local well pump and disease began to wane locally (Box 7.1). To his credit, Snow thought that the disease may have been declining before the pump handle was removed but he was correct in thinking that it was waterborne. It was another thirty years before Robert Koch obtained pure cultures of the causative organism, *Vibrio cholerae*.

Many bacterial pathogens are passive contaminants of environmental and drinking water; that is, such water is not their natural habitat and they do not actively replicate in it. *Vibrio cholerae* is an exception. It thrives in coastal waters and estuaries and will grow in water of low salinity that is warm and contains plenty of organic nutrients. Chitin is a key component of the exoskeletons of insects and shellfish and *Vibrio cholerae* is unusual in that it produces a chitinase that enables it to use this material as a source of carbon and nitrogen. More important, chitin induces genetic competence in *Vibrio cholerae*; that is, the bacterium acquires the ability to take up exogenous DNA (bacterial transformation) and this has played a key part in the development of pathogenicity.

Microbiology of Infectious Disease. Sandy B. Primrose, Oxford University Press.
© Sandy B. Primrose (2022). DOI: 10.1093/oso/9780192863843.003.0007

Figure 7.1. A cholera hospital in Bangladesh. Note that the empty beds are made up with plastic sheets that have a hole for drainage so that the large volumes of liquid stools can be collected in buckets.

Source: Mark Knobil/Flickr courtest of Wikimedia Commons. Reproduced under Creative Commons Attribution 2.0 Generic license. <https://commons.wikimedia.org/wiki/File:Cholera_hospital_in_Dhaka.jpg>.

The structure of the cell surface lipopolysaccharide of *Vibrio cholerae* is used to classify the bacterium into more than 100 serotypes. Until 1992 only one of these serotypes, O1, was known to cause cholera epidemics. This begs the question, why is only one serotype pathogenic? The O1 serotype is further classified into two biotypes: classical and El Tor, the latter getting its name because it was first isolated in El Tor in Egypt in 1905. Traditionally, the two biotypes were distinguished according to the presence of a haemolysin and the presence of the acetoin fermentation pathway, both of which are characteristic of El Tor strains. Today, these phenotypic tests have been replaced with sequence analysis of particular stretches of DNA. The fifth (1881–1896) and sixth (1899–1923) cholera pandemics originated in the delta of the River Ganges and bacterial isolates from them have the classical biotype. The seventh pandemic (1961 and ongoing) originated in Makassar on the island of Sulawesi in Indonesia and is caused by the El Tor biotype.

In 1992, a cholera epidemic began in Madras and to the surprise of epidemiologists, the causative organism was a non-O1 strain. Indeed, it was a completely new serotype designated O139 and soon it was the predominant cause of cholera on the Indian sub-continent. It was not clear whether the O139 strain was a non-O1 serotype that had acquired pathogenicity or an O1 strain with altered surface antigens. Biochemical characterization showed

---

Box 7.1 JOHN SNOW AND THE BROAD STREET WATER PUMP

In 1854, Soho was a suburb of London and in August of that year it was hit hard by an outbreak of cholera. Within a radius of 250 yards of the water pump in Broad Street there were 500 deaths from cholera within 10 days. Dr John Snow made a street plan and marked on it where each death had occurred. He then investigated each case to determine if the victim drank water from the Broad Street pump. In most cases they did, strongly suggesting that the pump was the source of infection. Equally important, Snow investigated groups of people who did not get the disease. A workhouse with over 500 inmates had virtually no cases and as it had its own well, it did not use the Broad Street pump. Men working in a brewery on Broad Street remained free of cholera and Snow discovered that the brewery also had its own well. Two women from the fashionable suburb of Hampstead also died from cholera and their deaths initially puzzled Snow. Then he discovered that one of the women previously lived on Broad Street. She liked the taste of the water so much (!) that she had it delivered to her regularly. The last delivery was the day before she died. Despite Snow's pioneering work it was years before the authorities made any significant effort to provide clean drinking water. When they finally acted it was found that a cesspit was contaminating the water going to the Broad Street pump.

---

that the new serotype was close to the El Tor biotype rather than the classical biotype. The new strain also had virulence characteristics typical of O1 strains and not found in non-O1 strains. Finally, the O139 strain had a mucous capsule outside its cell wall that is characteristic of non-O1 strains but not O1 strains. Subsequent analysis showed that the O139 strain arose by transfer of a group of genes ('genomic island') encoding surface antigens from a non-O1 strain to an O1 strain. This may have occurred by transformation as described earlier. Many people in the Indian sub-continent have developed immunity to O1 strains and this may have driven the evolution of the O139 strain.

The genomes of many pathogenic and non-pathogenic strains of *Vibrio cholerae* have been sequenced and the results show that, unlike most other bacteria, they have two non-homologous chromosomes. In addition, the genomes of pathogenic strains have acquired several mobile genetic elements known as pathogenicity islands. All pathogenic strains contain *Vibrio* pathogenicity islands 1 (VPI-1) and 2 (VPI-2). All isolates from the seventh pandemic also contain *Vibrio* seventh pandemic island 1 (VSP-1) and 2 (VSP-2). Pathogenic strains also contain another mobile genetic element, a bacterial virus (bacteriophage) called CTXΦ that integrates into the bacterial genome. Classical and El Tor strains of *Vibrio cholerae* have different versions of this bacteriophage but both versions encode the cholera toxin. The mode of action of cholera toxin is shown in Figure 7.2.

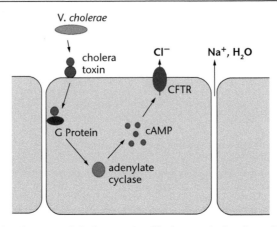

Figure 7.2. Mode of action of cholera toxin. Cholera toxin is released from bacteria in the gut lumen and binds via the B subunit to GM1 receptors on enterocytes, triggering endocytosis. Following activation in the cytosol of an infected cell, the A subunit enzymatically activates a G protein and locks it into its GTP-bound form through an ADP-ribosylation reaction. Cholera toxin can then go on to activate other G proteins. Constitutive G protein activity leads to activation of adenylyl cyclase and increased cAMP levels. High cAMP levels then go on to activate the membrane-bound CFTR protein, leading to dramatic efflux of chloride, sodium, and water from the intestinal epithelium.

Source: Reproduced with permission from David Lapierre/Sharing in Health <http://www.sharinginhealth.ca/pathogens/bacteria/species/vibrio_cholerae.html>.

VPI-1 is the master determinant of cholera pathogenesis because it encodes a structure on the surface of the bacteria known as a toxin co-regulated pilus (TCP) in addition to many of the virulence factors that contribute to the severity of disease. A pilus (plural: pili) is a tubular structure that projects from the surface of the cell. The TCP has two key roles in pathogenesis. First and foremost, it is the adsorption site for CTXΦ when this bacteriophage infects *Vibrio cholerae* cells. If there is no phage infection there will be no toxin production. Second, the TCP helps the pathogen to colonize the gastrointestinal tract. Thus, the presence of VPI-1 and CTXΦ are essential for *Vibrio cholerae* to be able to cause cholera. The other pathogenicity islands (VPI-2, VSP-1 and VSP-2) are not essential but when present they enhance pathogenicity.

An important defence mechanism employed by *Vibrio cholerae* is the type VI secretion system (T6SS). In the case of *Vibrio cholerae*, the T6SS enables it to evade predatory microfauna in the aquatic environment and to compete with other bacteria. The system might also allow the release of DNA from other bacteria that then can be taken up by transformation to give *Vibrio cholerae* new properties.

Why is *Vibrio cholerae* a 'surprising' pathogen as the heading of this chapter implies? Mention *Vibrio cholerae* to a microbiologist and they will immediately think of the disease cholera. But there are hundreds of serotypes of the organism living in aquatic habitats and, until the emergence of the O139 serotype in 1992, only the O1 serotype was a serious pathogen. We know that in order to cause epidemic disease in humans, *Vibrio cholerae* needs to carry the DNA of VPI-1 and CTXΦ. But if no other serotypes carry this DNA, then from where did the O1 serotype acquire it? Cholera toxin is an AB_5 toxin, consisting of one A subunit and five B subunits. The A subunit is catalytically active and is anchored to a ring of five B subunits which bind to epithelial cells. Another toxin with an identical structure is found in enterotoxigenic strains of *Escherichia coli* (p34), a bacterium completely unrelated to *Vibrio cholerae*. When the amino acid sequences of the two toxins are compared there is approximately 80% homology. The bacteriophage CTXΦ has a filamentous structure and resembles the filamentous phage M13 that infects *Escherichia coli*. Not only do both bacteriophages have genes in common but they both infect their hosts using pili. *Escherichia coli* inhabits the gut of animals and because it is shed in large numbers in faeces, it often finds its way into aquatic habitats. There it will encounter strains of *Vibrio cholerae* and in a very rare event, horizontal transmission may have occurred. If this is indeed the origin of the O1 strain, then changes to the *Escherichia coli* genes probably would have been needed. The origins of the seventh cholera pandemic provide a clue to the changes that might have occurred.

A large number of pilgrims travel to Mecca for the annual Haj. In the late nineteenth century, there was concern that cholera carried by these pilgrims could spread to neighbouring countries. Consequently, government monitoring facilities were established to monitor travellers entering Egypt and Iraq from Saudi Arabia to ensure that they were free from cholera. One such facility was at El Tor in Egypt and some of the strains of *Vibrio cholerae* that were isolated there have been retained to the present day. A pre-pandemic strain with the El Tor biotype was found around 1900. This contained VPI-1 but was not pathogenic because it lacked CTXΦ and so could not produce cholera toxin. Other pre-pandemic El Tor strains were isolated subsequently from other parts of the world and they have enabled us to piece together the many changes in the original El Tor strain that led to the seventh pandemic strain. These can be summarized as follows. Between 1903 and 1908 there was acquisition of CTXΦ but the resulting strain was a weak pathogen. Between 1908 and 1925, the strain migrated to Makassar in Indonesia, probably in a pilgrim from Mecca. In the same period, the strain acquired at least twenty-one mutations. Between 1939 and 1954 it acquired the mobile genetic elements VSP-1 and VSP-2 which enhance pathogenicity. More important, in the same period it acquired over fifty mutations. By the time the seventh pandemic started in 1961, there had been a further nine mutations and the causative strain is known as 7PET (seventh pandemic El Tor). Clearly, turning a harmless strain of *Vibrio cholerae* into a feared pathogen is very complex, which probably explains why it has only occurred once.

In the 1990s there was a cholera epidemic in South America that started in Peru but spread to Argentina. The latter country instituted the mandatory reporting of cholera cases and collected a very large number of samples. Recently, a large number of these samples were subjected to genomic analysis and the outcome was surprising. The vast majority of the isolates (425/490) were of the 7PET strain. The remaining strains were a mixture of O1 and non-O1 serotypes and appear to be endemic to Argentina. For some reason, these endemic strains do not spread epidemically. The presence of endemic strains of *Vibrio cholerae* is unlikely to be unique to Argentina. Rather, their existence was only detected by mass genome sequencing of isolates.

Key points

- The natural habitat of *Vibrio cholerae* is warm, nutrient-rich water, particularly estuarine water.
- *Vibrio cholerae* produces a chitinase that degrades the exoskeletons of insects and shellfish. The presence of chitin induces competence for genetic transformation that will permit horizontal gene transfer.
- Most strains of *Vibrio cholerae* do not cause disease but those that do mostly belong to the O1 serotype of which there are two biotypes: classical and El Tor.
- Disease-causing strains have at least two pathogenicity islands (VPI-1 and VPI-2) and carry the prophage CTXΦ. The prophage genome encodes cholera toxin.
- CTXΦ resemble coliphage M13 and the cholera toxin is similar to that found in enterotoxigenic *Escherichia coli*. Therefore, pathogenic strains of *Vibrio cholerae* may have acquired some of their virulence characteristics from *Escherichia coli* by horizontal gene transfer.
- The first six major epidemics of cholera were caused by the classical biotype of *Vibrio cholerae*. The seventh epidemic was caused by the El Tor biotype. The original El Tor isolates were of low virulence compared with the 7PET strain and the genetic changes that led to the creation of the 7PET strain have been identified.

Suggested Reading

Bik E.M., Bunschoten A.E., Gouw R.D., and Mooi F.R. (1995) Genesis of the novel epidemic *Vibrio cholerae* O139 strain: evidence for horizontal transfer of genes involved in polysaccharide synthesis. *The EMBO Journal* **14**, 209–16.

Dorman M.J., Domman D., Poklepovich T., Tolley C., Zolezzi G., et al. (2020) Genomics of the Argentinian cholera epidemic elucidate the contrasting dynamics of epidemic and endemic *Vibrio cholerae*. *Nature Communications* **9**, 2703. doi:10.1038/s41467-020-18647-7

Harris J.B., LaRocque R.C., Qadri F., Ryan E.T., and Calderwood S.B. (2012) Cholera. *The Lancet* **379**, 2466–76.

Hu D., Liu B., Feng L., Ding P., Guo X., et al. (2016) Origins of the current seventh cholera pandemic. *Proceedings of the National Academy of Sciences* **113**(48), E7730–E7739. doi:10.1073/pnas.1608732113

Kostiuk B., Unterweger D., Provenzano D., and Pukatzki S. (2017) T6SS intraspecific competition orchestrates *Vibrio cholerae* genotypic diversity. *International Microbiology* **20**, 130–7.

Kumar A., Das B., and Kumar N. (2020) Vibrio pathogenicity island-1: the master determinant of cholera pathogenesis. *Frontiers in Cellular and Infection Microbiology* **10**, 561296. doi:10.3389/fcimb.2020.561296

Mekalanos J.J., Rubin E.J., and Waldor M.K. (1997) Cholera: molecular basis for emergence and pathogenesis. *FEMS Immunology and Medical Microbiology* **18**, 241–8.

Spangler B.D. (1992) Structure and function of cholera toxin and the related Escherichia coli heat-labile enterotoxin. *Microbiology and Molecular Biology Reviews* **56**, 622–47.

8

The Accidental Pathogen: *Legionella pneumophila*

July 1976 was one of major historical significance in America: it was the bicentenary of the American Declaration of Independence. It was a time for celebration. However, it would be remembered for the wrong reasons by members of the American Legion who were attending a convention in Philadelphia. Just one day into the convention, people in attendance began to fall ill with symptoms not unlike those of pneumonia and by the end of the outbreak, twenty-nine members of the Legion had lost their lives. In total, 34 were confirmed to have died from the illness while 221 had become ill. After six months of intensive investigations, it was announced that the pneumonia-like symptoms had been caused by a newly discovered bacterium that was given the name *Legionella pneumophila*. However, the source of the bacterium remained a mystery although the sharply peaked epidemic curve suggested a common source.

Eight years before the outbreak of Legionnaires' disease in Philadelphia, there had been a mysterious outbreak of disease in Pontiac in Michigan among people who worked at and visited the city's health department. All the victims had fever and symptoms like those of influenza but there was no pneumonia. The cause of the disease was not known but it was given the name Pontiac fever. After the cause of Legionnaires' disease was identified, Michigan health officials tested blood samples that had been retained from people with Pontiac fever and discovered that they too had been infected with *Legionella pneumophila*.

Following the identification of *Legionella pneumophila* as the cause of Legionnaires' disease, other outbreaks of the disease were recorded, first in other US cities and then in Europe. The source of these outbreaks eventually was traced to water systems, particularly cooling towers associated with industrial plants and air-conditioning systems (Figure 8.1). The fans in the cooling towers generate aerosols that can spread up to six kilometres and, if *Legionella* is present in the water droplets, they get inhaled and can cause

Microbiology of Infectious Disease. Sandy B. Primrose, Oxford University Press.
© Sandy B. Primrose (2022). DOI: 10.1093/oso/9780192863843.003.0008

Box 8.1 THE PATHOGENESIS OF *LEGIONELLA PNEUMOPHILA* INFECTIONS

Droplets of water containing the bacteria are deposited in the lower respiratory tract where they are engulfed by alveolar macrophages, the primary defence against bacterial infection of the lungs. Instead of being killed by the macrophages, the bacteria multiply inside them and eventually the macrophages lyse releasing a new generation of *Legionella* to infect more cells. The death of the alveolar macrophages, plus the presence of bacteria, results in the production of signals (PAMPs) that attract monocytes and polymorphonuclear neutrophils. Leaky capillaries allow the slow release of serum and the deposition of fibrin in the alveoli resulting in an inflammation that obliterates air spaces and compromises respiratory function.

---

disease (Box 8.1). Other related sources of infection are the aerosols created by showers and hot tubs. The key to preventing Legionnaires' disease is to make sure that building water systems are maintained in such a way as to reduce the risk of *Legionella* growth and spread (Box 8.2) and this is a legal requirement in many countries. *Legionella* grows best in warm water but warm temperatures also make it hard to keep disinfectants, such as chlorine, at the levels needed to kill it.

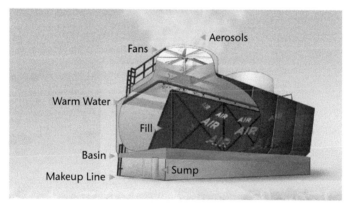

Figure 8.1. Cooling towers, which are often part of the air-conditioning systems of large buildings, are a common source of *Legionella* exposure in outbreaks. Cooling towers need to be properly maintained in order to prevent Legionnaires' disease.
Source: National Center for Immunization and Respiratory Diseases (NCIRD) (Public domain) <https://www.cdc.gov/legionella/images/materials-towers.jpg>.

That Legionnaires' disease exists at all is a consequence of modern building design because the natural habitat of *Legionella pneumophila* is freshwater environments. In such places, it coexists with other bacteria and microscopic eukaryotes such as amoebae, ciliates, and nematodes. Many of these microeukaryotes feed on bacteria and it might be thought that they have no preferences as to the bacteria they consume. This is not the case. When the amoebae *Acanthamoeba* and *Naegleria* were exposed to a mixture of *Escherichia coli* and *Legionella pneumophila*, they exhibited a preference for *Escherichia coli*. This was a wise choice. Once *Legionella pneumophila* is engulfed by protists, it inhibits their digestion process and begins to multiply inside them until it reaches high numbers (Figure 8.2). Within building water systems, *Legionella* colonizes existing multispecies biofilms and its intracellular habitat can even protect it from biocide treatments.

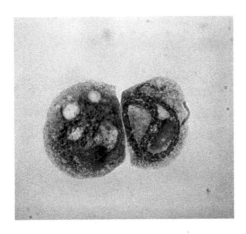

Figure 8.2. *Legionella pneumophila* (red chains) multiplying inside the protist *Tetrahymena pyriformis*.

Source: Don Howard/Centers for Disease Control and Prevention's Public Health Image Library (PHIL), ID #6632 (Public domain) <www.commons.wikimedia.org/wiki/File:T._pyriformis_hosting_L._pneumophila.png>.

What is the mechanism whereby *Legionella* escapes digestion? It has long been known that a bacterial type IV secretion system was essential, but what else? An answer has been provided by analysis of genomes from eighty different strains and species of *Legionella*. One surprise from this analysis has been the genomic diversity. The genome size varies from 2.37 million

---

Box 8.2 MANAGING *LEGIONELLA* IN WATER SYSTEMS

Legionella multiplies best in water at temperatures between 20 °C and 45 °C and which contains nutrients. To prevent growth of the bacterium in hot and cold water systems, hot water should be stored at 60 °C or higher and distributed at 50 °C or higher. Cold water should be stored and distributed below 20 °C. Stagnant water favours growth of *Legionella* and so dead legs/dead ends in pipework should be eliminated. Also, infrequently used showerheads and taps should be flushed weekly.

Legionella control in cooling water systems usually is done by the addition of biocides such as chlorine, bromine, and isothiazolinones. The effectiveness of these biocides can be adversely affected by the chemical composition of the water, its pH value, and the presence of corrosion and scale inhibitors. As these biocides degrade it is essential to ensure that a bactericidal concentration is present at all times.

base pairs to 4.88 million base pairs and the guanine + cytosine (GC) content from 34.8% to 50.9%. A decrease in GC content correlates with increasing genomic size and strongly suggests the addition of foreign DNA by horizontal gene transfer. Many bacterial pathogens that have an intracellular lifestyle undergo genome reduction (see, e.g., p63). In contrast, *Legionella* species appear to have undergone continuous genome expansion by horizontal gene transfer. Although nearly 18,000 genes have been identified, only 1,008 of them are common to all isolates. The genes involved in the T4SS are among those found in all isolates and the sequences of the proteins that they encode are highly conserved at the amino acid level, a strong indication of their essential nature.

Analysis of the potential genes in the different *Legionella* genomes resulted in the identification of more than 200 eukaryotic-like proteins and 137 proteins containing structural motifs normally found only in eukaryotes. A large number of these eukaryotic-like proteins are encoded by genes with a high GC content implying that *Legionella* acquired them by horizontal gene transfer from a eukaryotic donor. At least 180 of these proteins belong to a family of enzymes (GTPases) characterized by an ability to hydrolyse GTP to GDP, a key step in many eukaryotic cell signalling pathways. No counterparts of many of these GTPases are found in any other bacteria. One structural motif that has been found is the ergosterol reductase domain and this is of interest because ergosterol is a key component of the cell membranes of amoebae. These eukaryotic-like proteins and structural motifs allow *Legionella* to mimic host cellular processes as part of their replication strategy.

When an amoeba or a macrophage comes in contact with a bacterium it normally ingests it by the process known as phagocytosis. In this process the bacterium gets wrapped in host cell membrane into a structure known as a phagosome. This phagosome then fuses with a lysosome, a membrane-bound organelle found in macrophages and other cells that contain contains hydrolytic enzymes and toxic peroxides. Bacteria in such a phagolysosome are rapidly degraded. The events that occur with *Legionella pneumophila* are quite different. As soon as it comes in contact with a eukaryotic cell membrane it uses its T4SS to inject over 300 effectors into that cell. Many of these effectors are GTPases and they alter a whole series of intracellular processes such that the phagosome does not fuse with a lysosome. Rather, it develops a replicative niche known as a '*Legionella*-containing vacuole' that evades fusion with lysosomes and associates intimately with the host endoplasmic reticulum.

The bacteria multiply to a high number within the *Legionella*-containing vacuole (Figure 8.3) and this is facilitated by many of the effectors that they have secreted. Some of these effectors, such as the nucleomodulins, hijack

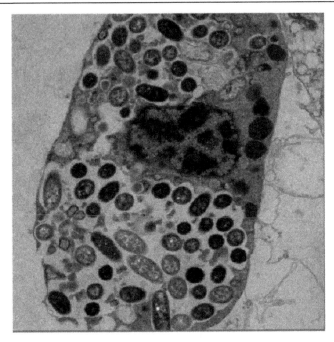

Figure 8.3. Transmission electron microscopy image of *Legionella pneumophila* within a phagocytic cell.

Source: TEM image of infected phagocytic cell (uncredited), reproduced under Creative Commons Attribution-Share Alike 3.0 Unported license <https://commons.wikimedia.org/wiki/File:TEM_image_of_Legionella_pneumophila_within_a_phagocytic_cell.tif>.

the cell nucleus and reprogramme gene expression to maximize the synthesis of proteins beneficial to bacterial growth and inhibit those that are detrimental. Other effectors interfere with ubiquitination, the eukaryotic process whereby proteins that are going to be degraded are marked by modification with ubiquitin. The key to the efficacy of these effectors is that they either resemble eukaryotic proteins or contain eukaryotic protein motifs. It is worth noting that the effector repertoire varies greatly between different *Legionella* strains and species and there is a very large functional redundancy. This redundancy is not necessary for infection of macrophages but reflects the need to avoid digestion by a wide array of amoeba and other eukaryotic predators in the natural environment. Indeed, it has been shown experimentally that only seven effectors are essential for growth of one strain of *Legionella pneumophila* in four different amoebae. All the other effectors affect the microeukaryote host range of the bacteria. However, some of the bacterial effectors that promote replication in amoebae restrict growth in macrophages and this could influence severity of disease caused by different strains.

In a sense, *Legionella pneumophila* is an accidental human pathogen. Humans seldom would have encountered the bacterium before the construction of modern buildings, which provide a breeding ground for *Legionella* and their host amoebae, and their associated aerosol-generating capabilities. Once humans did come in contact with the bacterium it is unfortunate that the mechanism whereby it evades digestion by amoebae also allows it to evade destruction by alveolar macrophages.

Key points

- *Legionella pneumophila*, the causative organism in Legionnaires' disease, is a natural inhabitant of freshwater environments. Unlike most aquatic bacteria which are digested by protists, *Legionella* multiplies inside protists after engulfment.

- Water distribution and air-conditioning systems in modern buildings have provided *Legionella pneumophila* with an ideal environment for growth. *Legionella* in aerosols from air-conditioning plants are inhaled by potential victims and deposited in the lung alveoli.

- *Legionella* escape digestion by alveolar macrophages in the same way that they escape digestion by protists in the environment. It does so by injecting over 300 effectors into the macrophages. Many of these effectors resemble eukaryotic proteins, or have motifs found only in eukaryotes, and enable *Legionella* to turn phagosomes into sites of replication.

Suggested Reading

Durie C.L., Sheedlo M.J., Min Chung J., Byrne B., Su M., et al. (2020) Structural analysis of the *Legionella pneumophila* Dot/Icm type IV secretion system core complex. *eLife* **9**, e59530. doi:10.7754/eLife.59530

Gomez-Valero L., Rusniok C., Carson D., Mondino S., Pérez-Cobas A.E., et al. (2019) More than 18,000 effectors in the *Legionella* genus genome provide multiple, independent combinations for replication in human cells. *Proceedings of the National Academy of Sciences* **116**, 2265–73.

Mondino S., Schmidt S., and Buchrieser C. (2020) Molecular mimicry: a paradigm of host–microbe coevolution illustrated by *Legionella. mBio* **11(5)**, e01201–20. doi:10.1128/mBio.01201-20

Mondino S., Schmidt S., Rolando M., Escoll P., Gomez-valero L., and Buchrieser C. (2020) Legionnaires' disease: state of the art knowledge of pathogenesis mechanisms of *Legionella. Annual Review of Pathology* **15**, 439–66.

Park J.M., Ghosh S., and O'Connor T.J. (2020) Combinatorial selection in amoebal hosts drives the evolution of the human pathogen *Legionella pneumophila*. *Nature Microbiology* **5**, 599–609.

Shaheen M. and Ashbolt N.J. (2021) Differential bacterial predation by free-living amoebae may result in blooms of *Legionella* in drinking water systems. *Microorganisms* **9**(1), 174. doi:10.3390/microorganisms9010174

Thomas D.R., Newton P., Lau N., and Newton H.J. (2020) Interfering with autophagy: the opposing strategies deployed by *Legionella pneumophila* and *Coxiella burnetii* effector proteins. *Frontiers in Cellular and Infection Microbiology* **10**, 599762. doi:10.3389/fcimb.2020.599762

9

Two Related Pathogens: One Ancient, One Modern

Neisseria is a large genus of bacteria that colonize the mucosal surfaces of many animals. Of the eleven species that colonize humans, the two closely related species *Neisseria meningitidis* (meningococcus) and *Neisseria gonorrhoeae* (gonococcus) are globally important causes of disease. The remaining nine species are non-pathogenic, except in immunocompromised hosts, but are a reservoir of genes for horizontal spread to the meningococcus and gonococcus. Phylogenetic analyses show that the meningococcus and gonococcus evolved from a common non-pathogenic ancestor but now are separate organisms that normally occupy distinct niches: the nasopharyngeal mucosa (meningococcus) and the genital mucosa (gonococcus). However, modern sexual practices have resulted in the meningococcus causing urethritis and the gonococcus causing oral and anal infections.

Meningitis is an acute inflammation of the protective membranes covering the brain and spinal cord, known collectively as the meninges, and usually occurs as a result of infection with viruses or bacteria. Three bacterial pathogens can cause meningitis: *Haemophilus influenzae*, *Streptococcus pneumoniae*, and *Neisseria meningitidis*. Of these, the meningococcus is the one that usually causes large outbreaks, particularly the epidemics seen in some parts of the world such as sub-Saharan Africa. Humans get infected with the meningococcus through contact with respiratory droplets or secretions that enter through the mouth from contact with a carrier. About 10% of humans are carriers where the bacteria have colonized the oropharynx and nasopharynx without causing clinical disease. For largely unknown reasons that are dependent on both the host and pathogen, in a small subset of carriers the meningococcus can invade the pharyngeal mucosal epithelium and, in the absence of bactericidal serum activity, disseminate into the bloodstream, thereby causing septicaemia. In a subset of cases, the bacteria

Microbiology of Infectious Disease. Sandy B. Primrose, Oxford University Press.
© Sandy B. Primrose (2022). DOI: 10.1093/oso/9780192863843.003.0009

can also cross the blood–brain barrier and infect the cerebrospinal fluid, causing meningitis (Figure 9.1).

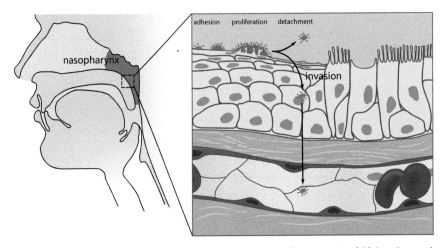

Figure 9.1. The human nasopharynx is the sole natural reservoir of *Neisseria meningitidis*. From there, bacteria colonize the mucosal surface as a commensal and can disseminate to the next host. They are sometimes found to cross the mucosal barrier and reach the underlying blood vessels, where meningococcal disease occurs.

Source: Arthur Charles-Orszag, reproduced with permission <https://www.researchgate.net/figure/Life-cycle-of-N-meningitidis-The-human-nasopharynx-is-the-sole-natural-reservoir-of_fig1_332961532>.

Gonorrhoea is a sexually transmitted disease caused by *Neisseria gonorrhoeae*. It has been known since ancient times and gonococcal DNA has been recovered from the dental calculus of Neanderthals. It is believed that this is the disease mentioned in the Book of Leviticus in the Old Testament where there is a reference to 'an issue of seed'. The name gonorrhoea was coined by the Greek physician Galen (131–200 AD) to describe 'an unwanted discharge of semen' (gono = seed, rhea = flow). We now know that the discharge they were describing was not semen but pus. The gonococcus is an obligate human pathogen as it cannot survive outside the human host. It usually infects the cervix and urethra but can also infect the pharynx and the rectum. It generally causes a non-complicated mucosal infection but occasionally can cause serious diseases in women such as inflammation of the fallopian tubes and pelvic inflammatory disease. Maternal transmission to children during birth can lead to neonatal blindness. The gonococcus manipulates the host immune response such that no immune memory is generated, thereby allowing repeat infections.

Box 9.1 PHASE VARIATION

Phase variation is a reversible process that helps bacteria to generate diversity quickly and adapt to rapidly changing environments and it plays a key role in the virulence of *Neisseria*. One mutation in a 'phase-variable' gene can switch expression ON, and another mutation in the same part of the DNA will switch the expression of the gene back OFF. Phase variation usually involves simple sequence repeats which are DNA tracts in which a short base pair motif is repeated several times. When DNA polymerase encounters simple sequence repeats during replication it can make copying errors. Addition or deletion of repeats can confer changes in protein sequence through frameshifts in open reading frames. Alternatively, it can alter the level of gene expression if the repeat sequence is in the promoter or regulatory region of a gene. Rates of phase variation often are several orders of magnitude greater than basal mutation rates and the frequency of switching is partly determined by the number of repeat units in any particular repeated sequence.

The genomes of the meningococci and gonococci are not large: they are ~ 2.1 Mb in size and encode about 2000 genes. The two genomes have about 68% homology. Both organisms share two properties that are important for their success as pathogens: both use phase variation of antigens (Box 9.1) to escape host defences and both are naturally transformable which facilitates horizontal gene transfer. *Neisseria meningitidis* has at least fifty phase-variable genes including ones involved in capsule biosynthesis, iron acquisition, and glycosylation of surface-expressed proteins. *Neisseria gonorrhoeae* has about seventy genes that are phase-variable and these are involved in surface antigen expression such as pili and the opacity-associated protein (Opa). Both neisserial pathogens have three phase-variable DNA methyltransferases which lead to variable genome-wide methylation differences and altered expression of multiple genes through epigenetic mechanisms.

Genetic transformation is the process by which a recipient bacterial cell takes up DNA from a neighbouring cell and integrates this DNA into its genome by recombination. In *Neisseria meningitidis* and *Neisseria gonorrhoeae*, DNA transformation requires the presence of short DNA sequences, 9–10 bases in length and known as DNA uptake sequences. These reside in coding regions of the donor DNA. The meningococcal genome contains about 2,000 copies of DNA uptake sequences indicating a high degree of horizontal gene transfer. Specific recognition of DNA uptake sequences is mediated by a type IV pilin. There is evidence of extensive gene transfer between all the human *Neisseria* strains, both pathogens and commensals. However, the meningococcus is believed to have acquired the ability to synthesize a capsule by horizontal gene transfer from the unrelated bacterium *Haemophilus influenzae* which also can colonize the nasopharynx.

One significant difference between the meningococcus and the gonococcus is that, of the two, only the meningococcus synthesizes a polysaccharide capsule. The capsule confers resistance to complement-mediated killing and phagocytosis. Whereas bacteria isolated from patients with meningococcal disease are capsulated, those from healthy individuals are not. Absence or decreased expression of a capsule usually is the result of downregulation of capsule gene regulation or changes in the capsule genes due to phase variation. Because the capsule is a key virulence component of meningococci it has been used in the development of vaccines against meningitis (Box 9.2).

Box 9.2 MENINGOCOCCAL VACCINES AND REVERSE VACCINOLOGY

The capsule of *Neisseria meningitidis* is a key virulence factor as it protects the bacterium from complement-mediated phagocytosis. There are twelve serogroups of the meningococcus and these are defined on the basis of the structure of the capsular polysaccharide. Six of these serogroups (A, B, C, W, X, Y) are responsible for most cases of disease worldwide. Effective vaccines for serogroups A, C, W, and Y have been available for many years and are based on making the relevant capsular polysaccharide immunogenic by conjugating it to diphtheria toxoid. There is no vaccine on the market for serogroup X but one based on a capsular polysaccharide conjugate is in clinical trials. Polysaccharide vaccines for serogroup B have not been found to be effective because of very poor immunogenicity of the capsule.

In the search for a serogroup B vaccine, scientists made use of a technique known as 'reverse vaccinology'. This is a genome-based approach. In comparison with the conventional approach, which requires a laborious process of selection of individual components important for immunity, reverse vaccinology offers the possibility of using genomic information derived from *in silico* analysis of sequenced genomes. Basically, the predicted open reading frames (ORFs) of the organism in question (in this case group B meningococci) are analysed to determine those that are expressed *in vivo* during infection. Once potential candidates are identified, their abundance is determined using proteomics and then a subset is subjected to immunogenicity determination. Two serogroup B vaccines have been developed using this reverse vaccinology approach. Bexsero (see Figure 9.2) comprises a mixture of a neisserial adhesin (NadA), the Neisserial heparin-binding antigen (NHBA), porin A (PorA), and factor H-binding protein (fHbp). The other vaccine, Trumenba, contains two subfamilies of fHbp and is used to prevent invasive meningococcal disease caused by *Neisseria meningitidis* serogroup B in individuals ten through twenty-five years of age.

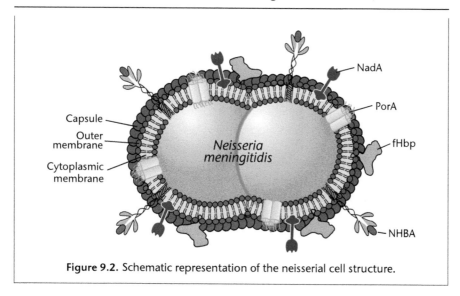

Figure 9.2. Schematic representation of the neisserial cell structure.

The natural habitat of *Neisseria meningitidis* is the nasopharynx where it grows on the top of mucus-producing epithelial cells surrounded by other microbes. The mucus is rich in antimicrobial peptides, components of the complement system, and immunoglobulin A (IgA) but the meningococcus is well equipped to survive in this environment. It expresses a protease that cleaves IgA and an efflux pump that prevents intracellular accumulation of antimicrobial peptides. It also produces a lipoprotein, factor H-binding protein (fHbp), which recruits human factor H (fH) to the meningococcal surface. This prevents complement from binding to the bacterium thus inhibiting bacteriolysis by the alternative complement pathway. The meningococcus competes with other bacteria in the mucus by producing antibacterial polymorphic toxins. The genes encoding these toxins are always accompanied by another gene encoding a protective immunity protein. This immunity protein protects the cell both from autointoxication and from toxins produced by other bacteria.

The mechanism whereby meningococci leave the nasopharynx and invade the bloodstream is not known. Three factors that may be important are NadA, LOS, and the MDA bacteriophage. NadA is a protein and LOS is a surface lipooligosaccharide that mediate adhesion to and invasion of epithelial cells. The Meningococcal Disease Associated (MDA) Island is a pathogenicity island that encodes the filamentous bacteriophage MDA. Phages are secreted and form a mesh of long filaments that bind the bacteria together and to epithelial cells. It is believed that the phages increase the bacterial load at the site of entry and enhance the probability of translocation into the bloodstream. Once in the bloodstream, meningococci use the same virulence factors as they do for survival in mucus. One virulence factor that is specific to bloodstream infections is the type IV pilus. This enables the bacteria to adhere to endothelial cells (cells lining blood vessels) that otherwise would not occur because of the presence of the capsule. The ability to stick to endothelial cells, a property almost exclusive to *Neisseria meningitidis*, probably facilitates the crossing of the blood–brain barrier.

Neisseria gonorrhoeae can survive as an intracellular organism within a variety of different cell types or extracellularly. Which state the organism enters depends largely on which surface components are expressed and whether these surface components are modified by glycosylation or phase variation. When gonococci reside extracellularly, they have pili which play a role in attachment to mucosal surfaces as well as to neutrophils. The latter are attracted to sites of infection by the release of proinflammatory cytokines and they are the major component of the characteristic gonorrhoeal discharge. The lipooligosaccharide (LOS) on the outer surface gets heavily modified with sialic acid residues (glycosylation) and this impedes the killing effect of serum. Internalization of the gonococcus requires loss of pili and loss of glycosylation of LOS. Other proteins that are required for internalization are PorA and Opa. PorA is a porin and it accounts for ~ 60% of the total protein of the gonococcus. Its role in infection is not fully understood but its level of expression points to its importance. Opacity-associated proteins (Opa) are integral outer membrane proteins that bind to various receptors on human cells.

Attempts to develop a vaccine against the gonococcus have not been successful. The phase variation of surface molecules makes them unsuitable as vaccine candidates which is the reason that most meningococcal vaccines make use of the capsular polysaccharide. Unfortunately, the gonococcus does not have a capsule. However, there is hope. Several studies have shown that the meningococcal group B vaccine Bexsero provides some cross-immunity to the gonococcus. In the absence of a vaccine, the clinical option is treatment rather than prevention. Unfortunately, *Neisseria gonorrhoeae* has developed or acquired resistance to all the antibiotics used against it including some recently introduced ones such as extended-spectrum cephalosporins and azithromycin. Individual antimicrobial determinants have evolved through selection or acquisition of chromosomal mutations and/or plasmids. Acquisition of the latter is facilitated because the gonococcus is naturally transformable throughout its life cycle.

Neisseria meningitidis and *Neisseria gonorrhoeae* share many pathogenicity/virulence determinants. Both bacteria inhabit mucosal surfaces but whereas the gonococcus can grow intracellularly, the meningococcus passes through cells and enters the bloodstream. However, the gonococcus existed in Neanderthals whereas the meningococcus emerged only about 400 years ago from a gonococcal ancestor. What drove this divergence? Currently we do not know.

Key points

- *Neisseria meningitidis* and *Neisseria gonorrhoeae* are the only *Neisseria* species to cause infections in immune competent individuals. The gonococcus has existed for thousands of years but the meningococcus only emerged about 400 years ago.
- Both bacteria use phase variation of surface antigens to escape host defences.
- Both bacteria are naturally competent for transformation and the large number of DNA uptake sequences in their genomes indicate that they have acquired a lot of genes by horizontal gene transfer.
- A key virulence determinant of the meningococcus is the capsule and the ability to synthesize it was acquired from the unrelated *Haemophilus influenzae*. Because of its importance, the capsule has been used in the development of vaccines. Meningococci produce many virulence factors but how these help the bacteria to invade the bloodstream is not known.
- Gonococci can exist both extracellularly and intracellularly. When residing extracellularly the gonococci attract neutrophils and these constitute most of the characteristic discharge seen in patients with gonorrhoea. Numerous virulence factors prevent destruction of the bacteria on the mucosal surface. The PorA protein probably plays a key role in internalization.

Suggested Reading

Caugant D.A. and Brynildsrud O.B. (2020) *Neisseria meningitidis*: using genomics to understand diversity, evolution and pathogenesis. *Nature Reviews Microbiology* **18**(2), 84–96. doi:10.1038/s41579-019-0282-6

Coureuil M., Jamet A., Bille E., Lécuyer H., Bourdoulous S., and Nassif X. (2019) Molecular interactions between *Neisseria meningitidis* and its human host. *Cellular Microbiology* **21**(11), e13063. doi:10.1111/cmi.13063

Hill S.A., Masters T.L., and Wachter J. (2016) Gonorrhoea—an evolving disease of the new millennium. *Microbial Cell* **3**, 371–89.

Quillin S.J. and Seifert H.S. (2018) *Neisseria gonorrhoeae* host-adaptation and pathogenesis. *Nature Reviews Microbiology* **16**, 226–40.

Wanford J.J., Holmes J.C., Bayliss C.D., and Green L.R. (2020) Meningococcal core and accessory phasomes vary by clonal complex. *Microbial Genomics* **6**(5), e000367. doi:10.1099/mgen.0.000367

10

Helicobacter pylori **and Gastric Ulcers**

For much of the twentieth century it was thought that gastritis (inflammation of the stomach lining) and peptic ulcers (open sores in the stomach
or duodenum) were caused by the stomach producing too much acid in
response to eating certain foods. The treatment was regular dosing with
medicines containing bicarbonate of soda coupled with avoidance of certain foods, particularly fried and spicy food. This treatment regime was not
particularly successful. A therapeutic breakthrough came in the late 1970s
with the development of cimetidine, an H_2 receptor antagonist that blocks
the production of stomach acid. This drug became a best-seller and soon was
followed by even better versions such as ranitidine and famotidine.

Just as cimetidine was coming to the market, an Australian pathologist
called Robin Warren had noted that gastric biopsy samples from patients
with gastritis had an unexpected burden of curved bacteria. In 1981, gas-
troenterologist Barry Marshall joined Warren in an attempt to culture the
bacteria from gastric samples and eventually they isolated the bacteria that
now are known as *Helicobacter pylori*. Their results eventually were published
in the prestigious journal *The Lancet*, but the journal editors had found it
difficult to find reviewers who considered the results to be significant. Conventional wisdom was that bacteria could not grow in the stomach because
of the low pH and clinicians were of the view that Warren and Mitchell were
totally wrong.

Frustrated by the failure of the medical fraternity to accept that *Helicobacter pylori* was the cause of gastritis and peptic ulcers, Marshall drank a pure
culture of the organism after having a baseline endoscopy. Soon he had
nausea and halitosis and by day 5 he began vomiting. Endoscopy on day
8 showed excessive gastritis and a gastric biopsy revealed that the bacterium
had colonized his stomach (Figure 10.1). He only recovered from the gastritis when he began to take antibiotics. This somewhat foolhardy experiment
confirmed that *Helicobacter pylori* causes gastritis as Koch's postulates had
been fulfilled. Even after details of this experiment were published, many

Microbiology of Infectious Disease. Sandy B. Primrose, Oxford University Press.
© Sandy B. Primrose (2022). DOI: 10.1093/oso/9780192863843.003.0010

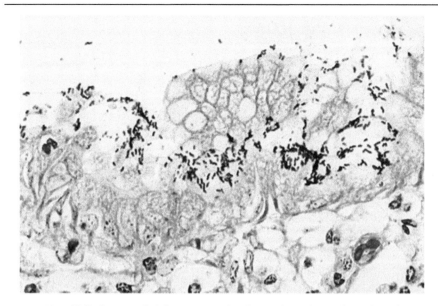

Figure 10.1. *Helicobacter pylori* (brown stain) colonized on the surface of epithelium.
Source: Yutaka Tsutsumi, M.D. Professor, Department of Pathology, Fujita Health University School of Medicine, Japan (Copyright free usage). <https://commons.wikimedia.org/wiki/File: Pylorigastritis.jpg>.

clinicians were unconvinced (see Box 10.1) and it was almost another ten years before the disease-causing potential of *Helicobacter pylori* was accepted. Today, it is known that in addition to causing gastro-duodenal ulcers, *Helicobacter pylori* infection can result in gastric carcinoma and lymphoma of the mucosa-associated lymphoid tissue (MALT) making *Helicobacter pylori* the only bacterium known to cause cancer. The persistence of Warren and Marshall paid off when they were awarded the 2005 Nobel Prize in Physiology and Medicine.

Any organism growing at the extremes of pH must have some mechanism for maintaining the intracellular pH near neutrality. *Helicobacter pylori* achieves this by producing the enzyme urease which breaks down urea into two buffering compounds: ammonia and bicarbonate. Urease can account for up to 10% of the soluble protein in the bacterium and is unusual in requiring nickel for activity. At neutral pH the enzyme contains little nickel and has minimal activity. As the pH value drops, the enzyme becomes increasingly loaded with nickel and its activity increases until at full activity each urease molecule has bound twenty-four nickel ions. As nickel is present at extremely low levels in normal diets,

---

Box 10.1 A PERSONAL ACCOUNT

In the early 1980s I was heading up the biotechnology function of an American pharmaceutical company. It had a new drug for treating gastritis and stomach ulcers in clinical trials which worked by increasing the production of acid-protectant mucus in the stomach. There was concern in the company that the work on *Heliobacter pylori* in Australia might negate the utility of their new drug. The worldwide Head of R&D asked me to review the Australian data. When I said that I believed the data, the clinical microbiologists in the company laughed at me: bacteria could not grow at a pH value of 2. However, I had been trained as a general microbiologist and knew of the existence of acidophilic bacteria, particularly those that occur in acid mine waters with a pH value below 1. Also, a few years earlier, as a result of a botched experiment, I had discovered that the well-studied bacterium *Escherichia coli* could drop the pH value of culture media to 2 and survive.

---

the bacterium synthesizes two histidine-rich proteins that are very effective at scavenging nickel. The ammonia produced by urease not only counteracts the stomach acid, it also disrupts tight cell junctions and thus damages the gastric epithelium. The other by-product, bicarbonate, is converted to carbon dioxide and this protects the bacterium from the bactericidal activity of nitric oxide and phagocytosis.

To cause disease, *Helicobacter pylori* needs to penetrate the mucous layer lining the stomach and colonize the gastric mucosa. The bacterium is spiral-shaped and it swims in a corkscrew fashion through the mucus. Passage is facilitated by the rise in pH caused by urease as this turns the mucus to liquid. Once through the mucus, the presence of outer membrane proteins and adhesins (e.g. OipA) enable adherence of *Helicobacter* to the gastric epithelial cells. The bacterium has a pathogenicity island that encodes a type IV secretion system and the cytotoxin-associated gene A (*cagA*). Other pathogenicity determinants are the vacuolating cytotoxin gene A (*vacA*) and duodenal ulcer promoting gene A (*dupA*) and outer membrane proteins such as OipA.

Once inside the host cell, the *cagA* gene product is localized on the plasma membrane where it is phosphorylated at specific Glu-Pro-Ile-Tyr-Ala (EPIYA) motifs by host kinases. The biological activity of *cagA*, and hence the degree of virulence, depends on the number and types of the EPIYA motifs. *CagA* interacts with multiple host cell molecules and is responsible for dysregulation of homeostatic signal transduction of gastric epithelial cells, induction of pro-inflammatory responses that lead to chronic inflammation of gastric mucosa, and induction of carcinogenesis (Figure 10.2). The VacA protein plays several roles in cellular pathogenicity but a key one is induction of cell necrosis. Most *Helicobacter pylori* strains carry a *vacA* gene, but there is marked variation among strains in toxin activity. This variation is attributable to strain-specific variations in amino acid sequences, as well as variations in the levels of gene expression and protein secretion.

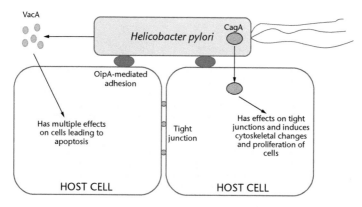

Figure 10.2. Mechanism of *Helicobacter pylori* induced host cell injury.

Helicobacter pylori transmission is believed to occur within families and acquired during childhood but the exact mechanism of transmission is not known. The most likely modes of transmission are the oral–oral and faecal–oral routes. Regardless of the mode of transmission, the bacterium infects approximately 50% of all humans although the infection rates vary from low in North America and Europe to very high in less developed parts of the world. Most of those infected will have no discernible symptoms because they carry strains of low virulence but ~ 15% will develop peptic ulcers and 1% will develop cancer. Thousands of strains of *Helicobacter pylori* have been collected from diverse geographical sources and distinct human populations such as the San in South Africa, the Amerinds in the Americas, Baka pygmies in Cameroon, and indigenous Australians. Two things became clear when the DNA of these strains was analysed. First, humans have been infected with *Helicobacter pylori* for at least 100,000 years, pre-dating the move of humans out of Africa ~ 60,000 years ago. The source of the initial infection remains unknown. Second, there are multiple, distinct populations that correlate with geographic origin. The genetic diversity in these different populations decreases with distance from sub-Saharan Africa, suggesting that *Helicobacter pylori* accompanied humans on their migration out of Africa. Looking at the high-level genetic structure it is possible to map the route that the bacterium took (Figure 10.3) and it closely matches the routes derived from analysis of human DNA (Box 10.2). One difference between the analyses of bacterial and human DNA is that the former suggests a second migration out of Africa about 52,000 years ago.

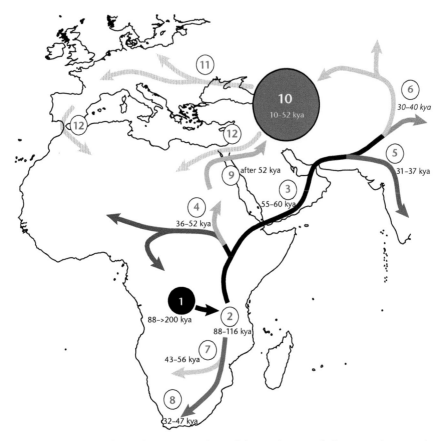

Figure 10.3. Chronological reconstruction of the major population events occurring during the intimate human-*H. pylori* association. Black lines indicate undifferentiated populations and all other lines are colour-coded according to population. The sequence of events is as follows: 1) Initial acquisition of *H. pylori* by a human ancestor; 2) Divergence of *H. pylori* into two super-lineages; 3) First successful migration of modern humans out of Africa via the southern route; 4) *H. pylori* divergence into hpAfrica1 and hpNEAfrica with migration eastwards (hpNEAfrica) and westwards (hpAfrica1); 5) Divergence of *H. pylori* out of Africa into hpSahul [8] and 6) hpAsia2 and hpEastAsia; 7) Host jump from San people to large felines giving rise to *H. acinonychis*. 8) Southward migration of San carrying the ancestor of hpAfrica2; 9) Second successful migration out of Africa via the Levant; 10) Hybridization of AE1 from central and south-west Asia and AE2 from north-east Africa in the Middle East or western Asia resulting in hpEurope; 11) Spread of hpEurope bacteria to Europe; 12) Back migration from the Middle East and Spain spreading hpEurope into North Africa. Figure reproduced by permission of *PLoS Pathogens* from Moodley et al. (2012).

Box 10.2   TRACING THE MIGRATION OF HUMANS OUT OF AFRICA

Mitochondrial DNA is present in all cells and is inherited via the maternal line. One part of the mitochondrial genome is particularly prone to mutation. By analysing the mutations in mitochondrial DNA from different populations, it is possible to trace the inheritance of modern humans back to their origins in Africa and their subsequent spread around the globe. The hypothetical woman at the root of all these changes is the matrilineal most recent common ancestor for all currently living humans and she is known as Mitochondrial Eve. The Y chromosome is inherited through the male line and mutational changes in its sequence also can be traced back to an origin in Africa. The maps of migration of mitochondrial and Y chromosome DNA are very similar and not unlike that shown in Figure 10.3.

The DNA of some bacterial species is preserved in teeth and bones (e.g. the plague bacillus, Chapter 3). However, *Helicobacter pylori* inhabits the stomach, an organ that is not usually preserved in ancient remains. One exception is mummification where soft tissues are preserved from post-mortem putrefaction. The first evidence for the existence of *Helicobacter* biomolecules in ancient remains came from the analysis of 1,700-year-old stool samples from a Chilean male. Analysis using ELISA (Enzyme Linked Immunosorbent Assay) revealed the presence of *Helicobacter* antigens. Another study used PCR amplification of 16S RNA and urease genes and confirmed the presence of *Helicobacter pylori* in a Mexican male dating to 1350 AD, in other words, before Columbus. Yet another study looked for the presence of *Helicobacter* DNA in an Amerindian mummy discovered in a glacier in Canada. Analysis revealed the presence of the *cagA* and *vacA* genes and their particular sequences was in accord with mitochondrial and Y-chromosome studies (Box 10.2) that suggest that the Americas were first colonized by migration over the Bering Strait. Genomic sequences for *cagA* and *vacA* also were found in the gut contents of Otzi the Iceman, a 5,300-year-old mummy that was recovered from a glacier on the border of Austria and Italy.

Key points

- Gastritis and peptic ulcers are caused by *Helicobacter pylori* and can be treated with antibiotics. *Helicobacter pylori* is the only bacterium that can cause cancer.

- To grow at the pH of the stomach, the bacterium produces urease that cleaves urea to ammonia and bicarbonate. The ammonia neutralizes the stomach acid and damages the gastric mucosa. The urease requires nickel for action and *Helicobacter* produces nickel-scavenging proteins.

- The increased pH liquefies the mucus layer and allows the bacteria to attach to the gastric epithelium where it uses a T4SS to inject effectors. The key effectors are CagA, which is pro-inflammatory, and VacA, which induces necrosis.

- Different populations of *Helicobacter pylori* exist in different parts of the world and analysis of these populations suggest that the bacterium evolved out of Africa alongside human migrations.

Suggested Reading

Ansari S. and Yamaoka Y. (2019) *Helicobacter pylori* virulence factors exploiting gastric colonization and its pathogenicity. *Toxins* **11**(1), 677. doi:10.3390/toxins11110677

Maixner F., Thorell K., Granehall L., Linz B., Moodley Y., et al. (2019) Helicobacter pylori in ancient human remains. *World Journal of Gastroenterology* **25**, 6289–98.

Moodley Y., Brunelli A., Ghirotto S., Klyubin A., Maady A.S., et al. (2021) *Helicobacter pylori's* historical journey through Siberia and the Americas. *Proceedings of the National Academy of Sciences* **118**(25), e2015523118. doi:10.1073/pnas.2015523118

Moodley Y., Linz B., Bond R.P., Nieuwoudt M., Soodyall H., Schlebusch C.M., et al. (2012) Age of the association between *Helicobacter pylori* and man. *PLoS Pathogens* **8**(5), e1002693. doi:10.1371/journal.ppat.1002693

Thorell K., Yahara K., Berthenet E., Lawson D.J., Mikhail J., et al. (2017). Rapid evolution of distinct *Helicobacter pylori* subpopulations in the Americas. *PLoS Genetics* **13**(2), e006546. doi:10.1371/journal.pgen.1006546

11

A Tale of Two Pathogens: *Pseudomonas aeruginosa* and *Pseudomonas syringae*

The genus *Pseudomonas* contains nearly 200 validly described species that demonstrate a great deal of metabolic diversity and consequently are able to colonize a wide range of niches. The best-known species is *Pseudomonas aeruginosa*, an opportunistic pathogen that is a major cause of nosocomial infections and that severely affects patients who are immunocompromised or suffering from cystic fibrosis. *Pseudomonas syringae* is the plant world's equivalent of *Pseudomonas aeruginosa* being an opportunistic pathogen of many different and important crops. The genomes of both organisms have been studied extensively and offer an opportunity to compare and contrast the disease process in plants and animals.

Pseudomonas aeruginosa

Pseudomonas aeruginosa can be found in a wide variety of environments, including soil and water, where it can utilize over 100 organic molecules as a source of carbon and energy. More important, it can be found readily in any environment occupied by humans and other animals and here it will exploit any weakness to cause disease. The bacterium is a common cause of bacterial keratitis, particularly in contact lens wearers, and often is associated with vegetation-related corneal insult (stick injuries). *Pseudomonas aeruginosa* is known to be particularly virulent in the eye and the keratitis it causes is known to be more difficult to treat and have worse prognosis than other forms of bacterial keratitis. The bacterium is one of the main causes of infections and sepsis in people suffering from severe burns and, as with infections in cystic fibrosis patients, its pathogenicity is correlated with its ability to form biofilms. It often is resistant to many classes of antibiotic, making it

Microbiology of Infectious Disease. Sandy B. Primrose, Oxford University Press.
© Sandy B. Primrose (2022). DOI: 10.1093/oso/9780192863843.003.0011

difficult to treat, and as a consequence it is listed as the highest priority for the development of new antibiotics by the World Health Organization.

The genome size of *Pseudomonas aeruginosa* varies greatly, ranging from 5.5 million to 7 million base pairs and encoding 5,500–6,600 genes. Analysis of the genome sequence of over 600 strains identified a pan-genome of 54,272 genes: 665 core genes present in all strains, 26,420 flexible genes found in many strains, and 27,187 genes unique to one particular strain. The diversity of genes in the pan-genome reflects the metabolic and ecological diversity of the bacterium. The smallest cellular genome so far identified is that of *Mycoplasma genitalium* and it is considered representative of the minimal set of genes required for a bacterium. The 665 core genes of *Pseudomonas aeruginosa* bear a striking resemblance to those found in *Mycoplasma genitalium*. However, the core gene set includes a number of genes responsible for pathogenesis that are not seen in *Pseudomonas* species that are not pathogenic for humans.

Pseudomonas aeruginosa produces a large number of virulence factors, too many to list here, along with a complex regulatory network of intracellular signals and cell–cell (e.g. quorum sensing, Box 11.1) signals. A major virulence determinant is the type III secretion system (T3SS) that injects effectors from the bacteria's cytosol directly into the cytoplasm of the eukaryotic host cell. Four key effectors are the exoenzymes ExoS, ExoT, ExoU, and ExoY. The majority of strains encode either ExoS or ExoU, but not both, and they facilitate distinctive mechanisms of bacterial propagation and pathogenesis. ExoS is associated with survival of *Pseudomonas aeruginosa* after it has been internalized through invagination of the host cell membrane. It blocks the formation of reactive oxygen species and alters the cell cytoskeleton resulting in inhibition of phagocytosis. ExoU is a phospholipase that mediates rapid destruction of the cell membrane of epithelial cells, macrophages, and neutrophils. Destruction of the latter leads to immunosuppressive effects that make the host more susceptible to infection.

Box 11.1 QUORUM SENSING

Quorum sensing is the ability of bacterial cells to communicate with each other and to regulate their behaviour in a density-dependent way. The 'language' used for this intercellular communication is based on small, self-generated signal molecules called autoinducers (AIs). The phenomenon of quorum sensing relies on the principle that when a single bacterium releases AIs into the environment, their concentration is too low to be detected. However, when sufficient bacteria are present, AI concentrations reach a threshold level that allows the bacteria to sense a critical cell mass and, in response, to activate or repress target genes (Figure 11.1). *Pseudomonas aeruginosa* has three quorum sensing systems and at least 10% of its genes are controlled by quorum sensing. Many of these genes are associated with virulence as shown by the fact that mutants with defective quorum sensing have greatly reduced virulence. Because of their association with virulence, quorum sensing systems are targets for the development of potential new antibiotics.

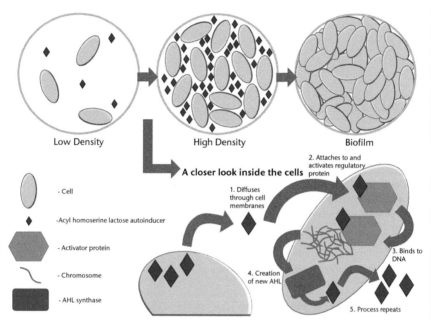

Low Density   High Density   Biofilm

A closer look inside the cells

- Cell
- Acyl homoserine lactose autoinducer
- Activator protein
- Chromosome
- AHL synthase

1. Diffuses through cell membranes
2. Attaches to and activates regulatory protein
3. Binds to DNA
4. Creation of new AHL
5. Process repeats

Figure 11.1. Cells excrete chemicals (autoinducers) into their environment to measure the population around it. Each type of bacterium has its own type of autoinducer and in Gram-negative bacteria it is most commonly an acyl homoserine lactone (AHL).

Source: Gluckma2/Wikimedia Commons, reproduced under Creative Commons Attribution-Share Alike 4.0 International license <https://commons.wikimedia.org/wiki/File:Quorum_sensing_of_Gram_Negative_cell.pdf>.

Many strains of *Pseudomonas aeruginosa* are lysogenized with multiple bacteriophages but the most widespread, and probably the most important, are the filamentous Pf ones of which there are at least eight. Unlike lytic bacteriophages, temperate phages such as Pf do not typically lyse their bacterial hosts. Rather, when filamentous phage virions are produced, they are extruded without bacterial lysis. Thus, in an infection caused by *Pseudomonas aeruginosa*, there also will be large numbers of free Pf bacteriophages. Anti viral immunity is triggered when leukocytes engulf Pf, and this results in decreased phagocytosis and reduced production of tumour necrosis factor— two factors required for the clearance of *Pseudomonas aeruginosa* infections. Thus, Pf bacteriophages are key virulence determinants.

Cystic fibrosis is a genetic disorder that principally affects the lungs which get colonized and infected by bacteria from an early age. These bacteria thrive in the altered mucus which collects in the small airways of the lungs (Figure 11.2). This mucus leads to the formation of bacterial microenvironments known as biofilms that are difficult for immune cells and antibiotics to penetrate. Over time, *Pseudomonas aeruginosa* begins to dominate in these biofilms and by eighteen years of age, 80% of patients with cystic fibrosis are infected with this bacterium. The formation of biofilm is induced and regulated by numerous genes and environmental factors and the most important of these are quorum sensing (Box 11.1), a molecule known as c-di-GMP, and Pf bacteriophages. The level of the signalling molecule c-di-GMP determines whether bacteria adopt either a planktonic or a biofilm phenotype.

Polymers present in the dense, viscous sputum that is characteristic of cystic fibrosis drive the self-assembly of free Pf bacteriophages into a highly ordered liquid crystal via depletion attraction. This is a cohesive force that operates between crowded, like-charged elements in environments where sufficient ionic strength exists to screen their repulsive forces. This crystalline architecture promotes the adhesiveness and viscosity of *Pseudomonas aeruginosa* biofilms. The high negative charge density of these structures

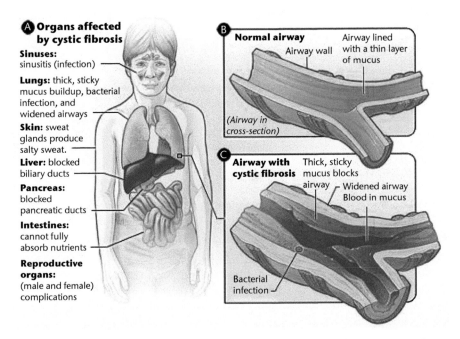

Figure 11.2. Figure A shows the organs that cystic fibrosis can affect. Figure B shows a cross-section of a normal airway. Figure C shows an airway with cystic fibrosis. The widened airway is blocked by thick, sticky mucus that contains blood and bacteria.

Source: National Heart Lung and Blood Institute (NIH), USA (Public domain) <https://commons.wikimedia.org/wiki/File:Cysticfibrosis01.jpg>.

makes them potent sequesters of cationic antibacterial agents, such as amino-glycoside antibiotics or host antimicrobial peptides. These effects are highly concentration dependent, with higher Pf bacteriophage concentrations promoting increasingly pathogenic *Pseudomonas aeruginosa* phenotypes.

For many years, the prevailing view was that individual cystic fibrosis patients acquired *Pseudomonas aeruginosa* infections separately and thus carried their own unrelated strains. However, in 1996 it was noticed that one particular strain had spread among patients in a children's cystic fibrosis centre in Liverpool in the United Kingdom This strain was dubbed the 'Liverpool Epidemic Strain' and it is associated with poorer patient health. Since then, other genetically distinct epidemic strains have been reported from Australia, Belgium, Canada, Denmark, and elsewhere in the United Kingdom. A key characteristic of epidemic strains is transmissibility, that is, the ability to transmit between patients, but the factors facilitating transmissibility are not known. However, epidemic strains are known to possess characteristic genomic islands.

Treatment of *Pseudomonas aeruginosa* infections has become a great challenge due to the ability of this bacterium to resist many of the currently available antibiotics. *Intrinsic* resistance of the bacterium includes low outer membrane permeability, expression of efflux pumps that expel antibiotics from the cell, and the production of antibiotic-inactivating enzymes. The *acquired* resistance of *Pseudomonas aeruginosa* can be achieved by either horizontal transfer of resistance genes (via plasmids, transposons, or bacteriophages) or mutational changes to antibiotic targets. Four different antibiotic efflux systems have been described: mexAB-oprM, mexXY-oprM, mexCD-oprJ, and mexEF-oprN. All classes of antibiotics except the polymyxins are susceptible to extrusion by one or more of these efflux systems. In addition, all *Pseudomonas aeruginosa* strains possess the *ampC* gene that encodes an inducible β-lactamase (p50) that destroys many penicillins and cephalosporins.

Pseudomonas syringae

Pseudomonas syringae is a bacterium that can grow *on* plants (epiphytic phase) or *in* plants (endophytic or pathogenic phase). There are many different pathovars of the bacterium and these cause a range of diseases on a diverse array of plants including annual crops (tomato, beans, etc.) as well as woody plants and trees (lilac, cherry, horse-chestnut, etc.). Symptoms include water-soaked lesions, blights, and cankers, and almost any part of the plant above ground can be infected. In the epiphytic phase, the bacteria live on the surface of leaves, stems, and fruits. Disease occurs when the bacteria gain access to the apoplastic compartment of plants through natural openings, such as stomata, or wounds, such as those caused by ice nucleation (Box 11.2). The apoplast is the extracellular space within plant structures and it contains proteins, metabolites, and inorganic ions that serve as nutrients for the pathogen.

Figure 11.3. Symptoms of brown spot on bean pods caused by *Pseudomonas syringae* pv. *syringae*. [Photo originally from USDA].

Source: Lyndon Porter, USDA ARS plant pathologist, reproduced with permission <https://mtvernon.wsu.edu/path_team/DiseaseGallery/bean-bac-brown-spot-5.htm>.

Box 11.2 ICE NUCLEATION

Although the melting temperature of ice is 0 °C, pure water does not freeze at that temperature. Rather, pure liquid water supercools to temperatures well below zero (–38 °C) before the phase transition to ice occurs. Ice formation can be triggered by the addition of nucleating agents and many organic and inorganic materials can fulfil this role. A number of bacteria that can be both saprophytes and pathogens (*Pseudomonas* sp., *Erwinia* sp., *Xanthomonas* sp.) also have the ability to promote ice nucleation. The genetic basis for ice nucleation in bacteria is known to be a single chromosomal gene, *ina*, which is both necessary and sufficient for the ice nucleation phenotype. This gene encodes an ice nucleation protein which is located on the outer surface of the producing bacteria and initiates ice formation by binding water molecules in a conformation resembling an ice lattice. The presence of ice nucleating bacteria on leaves and other plant parts prevents water on and within frost-sensitive plants from supercooling, thus leading to internal ice formation and frost damage.

Winter ski resorts often suffer from a lack of snow and generate 'artificial' snow by spraying water. As long as the temperature remains below about –8 °C, snow guns can readily produce as much snow as needed. When the thermometer reads higher, the water needs a little help to freeze and it is common to add *Pseudomonas syringae* to the water.

If *Pseudomonas syringae* encounters diseased or damaged tissue then it will begin the process of infection. However, healthy plants also are susceptible to infection if the right environmental conditions exist. A key driver of infection is high humidity and this acts in three ways. First, it increases the bacterial load on plant surfaces and there is a direct quantitative relationship between bacterial numbers and rates of infection. Second, it increases the swarming motility of the bacteria meaning that there will be more probing of the plant's defences. Third, a key entry point in healthy plants is via the stomata and the plant responds to bacterial signals by closing its stomata. However, high humidity suppresses this bacterial-induced stomatal closure.

Once *Pseudomonas syringae* gets into the apoplast a battle ensues between plant defences and bacterial virulence factors. There are two parts to the plant defences. First, the plant recognizes certain structures on the bacterial surface, the pathogen-associated molecular patterns (PAMPs), and develops pattern-triggered immunity. The bacteria respond by producing various virulence factors such as coronatine (Box 11.3) and various effectors that are delivered to the plant via a type III secretion system. Coronatine is a molecular mimic of jasmonate, a plant hormone that mediates plant defences (Box 11.3). Second, the plant mounts defences against the bacterial effector molecules via effector-triggered immunity. The result of these two plant defence systems is that the invading pathogen is inhibited from dividing but how this happens is not known.

One unifying feature of strains classified as *Pseudomonas syringae* is the presence of a type III secretion system. Multiple versions of this system exist but the most prevalent among strains whose genomes have been sequenced is referred to as the tripartite pathogenicity island because the genomic island containing it is arrayed into three distinct sections. This particular system is required for virulence *in planta* in every pathogenic strain investigated thus far, and its presence is strongly correlated with pathogenic potential on agriculturally relevant plants. A large number of different effectors have been identified in strains of *Pseudomonas syringae* but each strain has its own

---

Box 11.3 PLANT HORMONES AND PLANT DEFENCES

Jasmonic acid and salicylic acid are two key molecules for plant defences. Jasmonic acid protects plants against attack by herbivores whereas salicylic acid protects plants from bacterial pathogens by inducing expression of defence-promoting genes and antimicrobial pathogen-response genes. The two hormones effectively work in opposition: jasmonic acid signalling suppresses salicylic acid signalling and vice versa. Coronatine is a molecular mimic of jasmonic acid and its presence in cells results in a suppression of salicylic acid production making the plant cells more susceptible to *Pseudomonas syringae*.

repertoire and very few particular effectors are present in most strains. The effectors secreted into the apoplast by the T3SS do not act alone but operate in conjunction with other virulence factors, such as low molecular-weight toxins and pathogen-derived phytohormones, to suppress the host immune system and to modulate host physiology to favour infection.

Comparison of *Pseudomonas aeruginosa* and *Pseudomonas syringae*

Pseudomonas aeruginosa and *Pseudomonas syringae* have many phenotypic properties in common but they also have similar genome sizes. Although the size of the core genome has not been as extensively investigated in *Pseudomonas syringae*, it also has a very large number of flexible genes. Both bacteria are opportunistic pathogens. On the one hand they exist as skin commensals or plant epiphytes and, on the other, when the conditions are favourable, they can overcome host defences and cause disease. A key virulence determinant of both bacteria is the type III secretion system and although the effectors secreted are very different, most of the time they are fulfilling the same function. Both bacteria use quorum sensing systems to control the production of virulence factors and two of these should be highlighted. The first of these factors are the exoenzymes that destroy host tissue. The second is the production of an alginate polymer. This secreted polymer facilitates the production of lung biofilms and increases the viscosity of the mucus in sufferers of cystic fibrosis. It also is a key component of biofilms that increase the pathogenicity of *Pseudomonas syringae* for plants. One key difference between the two pathogens is that *Pseudomonas syringae* does not have an equivalent to the Pf bacteriophages that play a key role in the pathogenicity of *Pseudomonas aeruginosa* but overall, there are a lot of similarities between the two pathogens.

Key points

- *Pseudomonas aeruginosa* and *Pseudomonas syringae* are opportunistic pathogens, the former of animals and the latter of plants, that usually enter their hosts through wounds. They share many features relating to virulence.

- The two bacteria have a core genome that is similar in size to the postulated 'minimal bacterial genome'. The non-core genome is very large and reflects the metabolic and ecological diversity of the genus.

- Both species produce a wide range of virulence factors that are controlled by quorum sensing and many are injected into cells by a T3SS. Two virulence factors that are shared are the production of exoenzymes to destroy host tissue and the polysaccharide alginate that promotes biofilm formation. The latter plays a key role in *Pseudomonas aeruginosa* infections of patients with cystic fibrosis.

- The production of Pf bacteriophages is a key virulence determinant of *Pseudomonas aeruginosa*. There is no counterpart in *Pseudomonas syringae*.

Suggested Reading

Arnold D.L. and Preston G. (2019) *Pseudomonas syringae*: enterprising epiphyte and stealthy parasite. *Microbiology* **156**, 251–53. doi:10.1099/mic.0.000715

Dettman J.R. & Kassen R. (2020) Evolutionary genomics of niche-specific adaptation to the cystic fibrosis lung in *Pseudomonas aeruginosa*. *Molecular Biology and Evolution* **38**, 663–75.

Diggle S.P. and Whiteley M. (2019) Microbe profile: *Pseudomonas aeruginosa*: opportunistic pathogen and lab rat. *Microbiology* 116(1), 30–3. doi:10.1099/mic.0.000860

Freschi L., Vincent A.T., Jeukens J., Edmond-rheault J-G., Kukavica-Ibrulj I., et al. (2018) The *Pseudomonas aeruginosa* pan-genome provides new insights on its population structure, horizontal gene transfer, and pathogenicity. *Genome Biology and Evolution* **11**, 109–20.

Secor P.R., Burgener E.B., Kinnersley M., Jennings L.K., Roman-Cruz V., et al. (2020) Pf bacteriophage and their impact on *Pseudomonas* virulence, mammalian immunity, and chronic infections. *Frontiers in Immunology* **11**, 244. doi: 10.3389/fimmu.2020.00244

Wardell S.J.T., Rehman A., Martin L.W., Winstanley C., Patrick W.M., and Lamont I.L. (2019) A large-scale whole-genome comparison shows that experimental evolution in response to antibiotics predicts changes in naturally evolved clinical *Pseudomonas aeruginosa*. *Antimicrobial Agents and Chemotherapy* **63(12)**, e01619–19. doi:10.1128/AAC.01619-19

Xin X-F., Kvito B., and He S.Y. (2018) Pseudomonas syringae: what it takes to be a pathogen. *Nature Reviews in Microbiology* **16**, 316–28.

12

The Enigmatic Pathogens: Syphilis, Yaws, Pinta, and Bejel

Sexually transmitted diseases confer on their victims a social stigma that other diseases do not and syphilis is no exception. This is particularly interesting in the case of syphilis because the same organism, *Treponema pallidum*, causes three other diseases: yaws, bejel, and pinta. Yaws is a tropical infection of the skin, joints, and bones (Figure 12.1) and it is spread, usually non-sexually, by direct contact with the fluid from a lesion of an infected person. The disease is most common among children who play together. Bejel, or endemic syphilis, is a chronic infection of skin and tissue that can become invasive (Figure 12.2). Pinta is a much more minor disease that causes skin infections. The spiral bacteria (spirochetes) causing these four diseases (Figure 12.3) cannot be differentiated morphologically or by chemical or immunological methods but they are given the status of sub-species: *Treponema pallidum pallidum* (TPA, syphilis), *Treponema pallidum pertenue* (TPE, yaws), *Treponema pallidum endemicum* (TEN, bejel), and *Treponema carateum* (pinta). These pathogens share a genetic identity of over 99.7% and only subtle genetic differences have been identified that permit them to be distinguished. This begs the question of how such a limited variation can translate into the observed differences in pathogenesis.

The *Treponema pallidum* genome is ~ 1.14 million base pairs in size, one of the smallest among pathogenic bacteria. The genomes of over fifty strains have been completely sequenced and, unusually for a pathogen, they contain no prophages, insertion sequence elements, or plasmids. The human treponemes cannot be cultivated in the laboratory and genome sequencing has provided an explanation for their obligate pathogenicity. The genome has undergone reduction by dispensing with genes for synthesis

Microbiology of Infectious Disease. Sandy B. Primrose, Oxford University Press.
© Sandy B. Primrose (2022). DOI: 10.1093/oso/9780192863843.003.0012

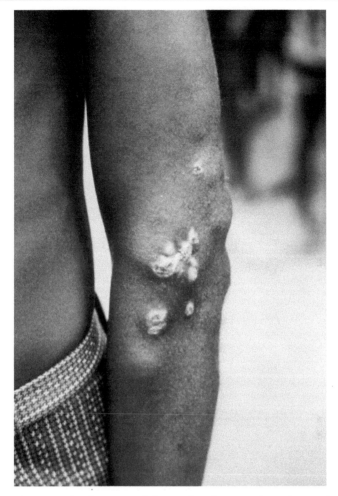

Figure 12.1. An example of yaws: nodules on the elbow resulting from infection with *Treponema pallidum pertenue.*

Source: CDC/Dr. Peter Perine (Public Health Image Library #3842), 1979 (Public domain) <https://commons.wikimedia.org/wiki/File:Yaws_01.jpg>.

of nucleotides, fatty acids, amino acids, and vitamins, as well as the ATP-generating capacity provided by the tricarboxylic acid cycle and oxidative phosphorylation. What treponemes need in the way of cellular building blocks they get from their hosts.

A comparison was made of the genomic sequences of fifty-three *Treponema pallidum* strains isolated from different parts of the world and a diverse range of hosts. This identified four pathogenicity islands and eight genomic islands

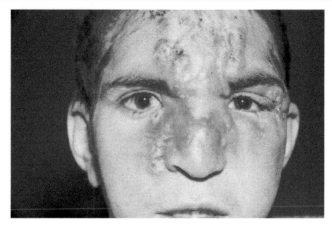

Figure 12.2. An example of bejel showing disfiguring infiltration of the nose and fore-head with clustered nodules caused by infection with *Treponema pallidum endemicum*.
Source: Journal of Emerging Infectious Diseases (Public domain) <http://wwwnc.cdc.gov/eid/article/19/1/12-0756-f1.htm>.

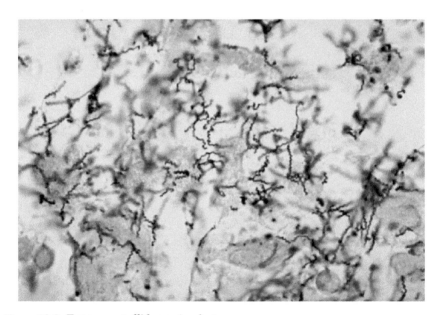

Figure 12.3. *Treponema pallidum* spirochetes.
Source: CDC/Dr Edwin P. Ewing, Jr (Public Health Image Library #836), 1986 (Public domain) <https://commons.wikimedia.org/wiki/File:Treponema_pallidum_01.png>.

in TPA (syphilis), three pathogenicity islands and seven genomic islands in TPE (yaws), and three pathogenicity islands and eight genomic islands in TEN (bejel). The pathogenicity islands were identical within a sub-species but not between them suggesting that they account for the differences in pathogenesis between the sub-species. Curiously, many of the genes carried by the pathogenicity islands have been identified but they do not encode proteins with obvious pathogenic potential. Consequently, how a treponeme causes disease is an enigma.

Antibiotic resistance plasmids usually carry genes that encode enzymes that can inactivate antibiotics. It is not known why treponemes do not carry any plasmids but the fact that they do not explains why penicillin still is effective after seventy years of use. The second antibiotic used to treat treponemes is azithromycin and resistance to it is becoming widespread. However, resistance in this case is not due to inactivation of the antibiotic but to a chromosomal mutation in the gene for its 23S ribosomal RNA target.

The emergence of HIV/ AIDS in the 1980s (p000) led to a major decline in the incidence of syphilis because people who engaged in casual sex adopted protective measures. However, the incidence of syphilis now is rising rapidly and there are over 6 million new cases every year, many of these in men who have sex with men or who indulge in sexual tourism. Much is known about the disease (Box 12.1) but one issue that has puzzled microbiologists for a long time is how the bacterium evades the host defence systems. Genomic sequencing may have provided the answer.

Box 12.1 SYPHILIS AS A DISEASE

Venereal syphilis is acquired when treponemes are inoculated onto the mucosa or skin during sexual contact. The treponemes directly penetrate mucous membranes or they enter through breaches in skin produced by sexual activity, and as few as ten organisms are required to cause an infection. Attachment to host cells and the extracellular matrix is considered to be the crucial initial step of infection. Once below the epithelium, the bacteria multiply locally and disseminate through the lymphatics and the bloodstream. The first sign of infection is an ulcerative lesion, the chancre, one to twelve weeks after the initial infection. This is readily noticed if it occurs on the penis but is easily missed if it is present on the cervix, throat, or rectum. By this stage the bacterium already is blood-borne. After a further four to ten weeks the symptoms of secondary syphilis appear as mucocutaneous lesions. The treponemal burden in blood and other tissues is highest at this stage and if the patient is a pregnant woman there is a high risk of transmission to the foetus. In about 40% of patients, the bacterium may invade the central nervous system and if this occurs it can result in serious neurological disorders in later years. If syphilis is left untreated it will enter a latent period when the patient is

continued

Box 12.1 *Continued*

---

free of obvious symptoms and is lulled into a false sense of security in thinking that they are free of the disease. They are not: it will recur. Whilst in the latent state, *Treponema pallidum pallidum* can hide from the body's defence systems but where it hides is not known and it has been dubbed the 'stealth pathogen'.

Early-stage syphilis is curable with a single intramuscular injection of benzathine penicillin, the drug of first choice for all stages of syphilis. The benzathine component slowly releases the penicillin making the combination long acting hence its efficacy with a single injection. The second drug used to treat the disease is azithromycin and this is preferred in underdeveloped countries because it can be administered orally. However, resistance to this antibiotic is widespread. Some historical treatments for syphilis are documented in Box 12.2.

In the treponemes there is a family of very closely related genes known as the *Treponema pallidum* repeat gene family (*tpr*) that encode major surface proteins. Based on sequence analysis of many different isolates from different individuals it is known that one of those genes, *tprK*, has seven variable regions (V1 to V7) separated by conserved amino acid sequences. When rabbits are infected with *Treponema pallidum*, variation in regions V1 to V7 accumulate over the course of infection. This antigenic variation could be the mechanism whereby the bacterium evades the host's immune response. Confirmation of this came from by genomic sequencing of blood-derived isolates collected six years apart from an individual who had syphilis four times in this period: there were virtually no changes at all across the entire genome except in one gene, *trpK*.

Box 12.2 SOME HISTORICAL TREATMENTS FOR SYPHILIS

Starting in the Middle Ages, the standard treatment for syphilis was mercury. It was administered in various ways, including by rubbing it on the skin, by applying a plaster, and by mouth! An even more dangerous method of administration was fumigation in which mercury was vapourized over a fire and the patients were exposed to the resulting steam, either by being placed in a bottomless seat over the hot coals, or by having their entire bodies except for the head enclosed in a box that received the steam.

A more rational treatment was pyrotherapy. It was noticed that patients who developed high fevers were cured of syphilis. Consequently, malaria was used as treatment for tertiary syphilis because it produced prolonged and high fevers. This was considered an acceptable risk because the malaria could later be treated with quinine. Julius Wagner-Jauregg won the 1927 Nobel Prize for Medicine for his discovery of the therapeutic value of malaria inoculation in the treatment of neurosyphilis.

A major breakthrough came with the development of the first modern antimicrobial agent, the organoarsenic compound arsphenamine. This was originally called '606' because it was the sixth in the sixth group of compounds synthesized for testing in the laboratory of Paul Ehrlich. It was marketed in 1910 under the trade name Salvarsan. Being an arsenic compound it had many side effects and was replaced with penicillin in the 1940s. More recently it has been shown to be a mixture of three compounds (Figure 12.4).

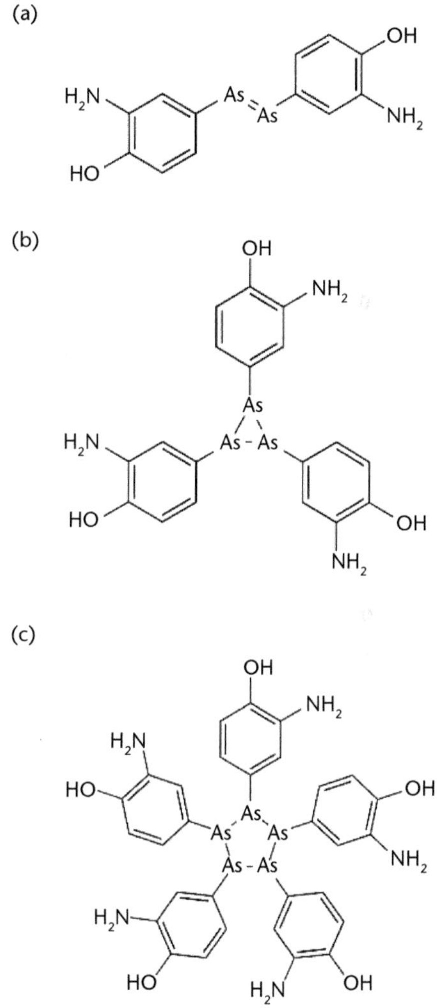

(a)

(b)

(c)

Figure 12.4. The three components of Salvarsan.

Source: Benjah-bmm27/Wikimedia Commons (Public domain) <https://commons.wikimedia.org/wiki/File:Salvarsan-montage.png>.

Yaws is much less common than syphilis with fewer than 500,000 new cases every year. However, because it is a disease of underdeveloped countries it often is not treated and about 2.5 million people are living with it. About 75% of the people with yaws are under fifteen years of age and they suffer from chronic disfigurement and disability. Between 1954 and 1962, the

World Health Organization (WHO) and UNICEF attempted to eradicate yaws globally by administering injectable penicillin in forty-six countries. This reduced the incidence by 95%. At this stage, responsibility for the elimination of the last 5% of cases passed from the global agencies to the healthcare systems of individual countries. Not surprisingly, most countries had other priorities and eradication failed.

In 2012, the WHO launched a fresh yaws eradication programme based on oral administration of azithromycin (see Box 12.1) to individuals in communities where the disease is endemic followed by clinical and serological surveys to detect any remaining cases. The programme is having success and in 2017, India declared itself free from the disease. However, there are potential pitfalls. First azithromycin resistance is beginning to emerge and its spread will need to be prevented by administration of penicillin. Second, bacteria similar to TPE have been found in baboons in Tanzania and it has been shown experimentally that they can infect humans. Non-biting flies are known to suck the secretions from the skin ulcerations of yaws patients and treponemal DNA sequences have been found in such flies. Thus, there is a chance that flies could transmit yaws from animal reservoirs such as baboons to individuals who have broken skin for any reason.

In the late fifteenth century, around the time of Christopher Columbus' discovery of the New World, there were accounts of an epidemic of a previously unrecorded disease that we now know as syphilis. The timing of these accounts has led some people to speculate that syphilis arose in the New World, possibly from yaws, and was brought to Europe by the sailors who accompanied Columbus. Other people believe that syphilis already was in Europe before the first voyage of Columbus and was taken to the New World by the early explorers. Physical examination of skeletal remains is of little value as the lesions on bones caused by syphilis are very similar to those caused by yaws.

Recently it has been shown that treponemal DNA can be recovered from skeletal remains and analysis of this DNA is beginning to shed light on the origins of syphilis. One study found TPP (yaws) DNA in the skeleton of a fifteenth-century plague victim in Lithuania. Given that yaws is a disease of tropical regions, this indicates that even in the fifteenth century there was movement of people between Africa and Northern Europe. Another study found TPA (syphilis) DNA in bones from neighbouring Estonia that were carbon dated to the mid-fifteenth century—before Columbus sailed to the New World. However, carbon dating does have methodological uncertainties and the bones could be from a later period. Perhaps the best evidence for syphilis in Europe before the voyages of Columbus comes from written records of the Italian War of 1494–1498. These describe an outbreak of syphilis caused by French troops during the invasion of Naples in 1495.

Key points

- *Treponema pallidum* causes four different diseases: syphilis, yaws, bejel and pinta. The only differences between the strains causing these diseases are in the pathogenicity islands that they have in their genome.

- Sequencing has shown that the *Treponema pallidum* genome has no insertion sequences, phages, or plasmids. Because of the absence of mobile genetic elements, the bacterium has not acquired resistance to the first-line antibiotic, penicillin. Resistance to the second-line drug, azithromycin, is developing because of mutations the target protein.

- *Treponema pallidum* is an obligate pathogen and genome sequencing has provided an explanation: the bacterium cannot synthesize nucleotides, amino acids, fatty acids, or vitamins, and has no tricarboxylic acid cycle.

- *Treponema pallidum* is able to escape immune detection and does this by changes in the seven variable regions of the *tprK* protein.

- The WHO is trying to eliminate yaws but resistance to azithromycin may make this target difficult.

- The question whether syphilis existed in Europe before the discovery of the Americas or was brought to Europe from the Americas is unresolved but on balance, probably pre-existed Columbus' voyages.

Suggested Reading

Addetia A., Tantalo L.C., Lin M.J., Xie H., Huang M-L., et al. (2020) Comparative genomics and full-length TprK profiling of *Treponema pallidum* subsp. *pallidum* reinfection. *PLOS Neglected Tropical Diseases* **14**(4), e0007921. doi:10.1371/journal.pntd.0007921

Griffen K., Lankapalli A.K., Sabin S., Spyrou M.A., Posth C., et al. (2020) A treponemal genome from a historic plague victim supports a recent emergence of yaws and its presence in 15th century Europe. *Scientific Reports* **10**, 9499. doi:10.1038/s41598-020-66012-x

Holmes A, Tildesley MJ, Solomon AW, Mabey D, Sokana O, et al. (2020) Modelling Treatment Strategies to Inform Yaws Eradication. *Emerging Infectious Diseases* **26**, 2685–93.

Jaiswal A.K., Tiwari S., Jamal S.B., de Castro Oliveira L., Alves L.G., et al. (2020) The pan-genome of *Treponema pallidum* reveals differences in genome plasticity between subspecies related to venereal and non-venereal syphilis. *BMC Genomics* **21**, 33. doi:10.1186/s12864-019-6430-6

Majander K., Pfrengle S., Kocher A., Neukamm J., du Plessis L., et al. (2020) Ancient bacterial genomes reveal a high diversity of *Treponema pallidum* strains in early modern Europe. *Current Biology* **30**, 3788–803.

Stamm L.V. (2016) Flies and yaws: molecular studies provide new insight. *EBioMedicine* **11**, 9–10.

13

Mycobacterial Mysteries: Tuberculosis and Leprosy

Members of the bacterial genus *Mycobacterium* are characterized by a unique cell wall structure (Figure 13.1) that is rich in long-chain fatty acids known as mycolic acids. These acids give the bacteria a waxy outer surface that is extremely hydrophobic and makes them resistant to many antibacterial agents. The Greek prefix 'myco' means 'fungus' and alludes to the fact that when mycobacteria are cultured on a solid medium, they have a mould-like appearance.

There are about 190 species of *Mycobacterium* but only a very small number are known to be pathogens. These pathogens can be classified into several major groups for the purpose of diagnosis and treatment. Members of the *Mycobacterium tuberculosis* complex (MTBC) can cause tuberculosis and they include *Mycobacterium tuberculosis*, *Mycobacterium bovis*, *Mycobacterium africanum*, and *Mycobacterium canetti*. The name 'tuberculosis' is derived from the characteristic tubercles that are formed in the lungs and other affected organs (Figure 13.2) Leprosy, or Hansen's disease, is caused by *Mycobacterium leprae* and *Mycobacterium lepromatosus*. The nontuberculous mycobacteria (NTM) are all the other pathogenic mycobacteria and they can cause pulmonary disease resembling tuberculosis, lymphadenitis, skin disease, or disseminated disease. The best known of these is *Mycobacterium ulcerans* that causes Buruli ulcer. The common feature of all the mycobacterial diseases is that the bacterium has found a way to defeat the host immune system resulting in chronic infections.

Leprosy is one of the most stigmatized diseases worldwide and originally was thought to be a punishment from God on sinful persons due to the fact that only certain people developed this disease. Leprosy is spread between people by contact and supposedly lepers had to warn others of their approach by ringing a bell and shouting 'unclean, unclean'. Also, lepers often were confined to leper colonies, a practice that continues to this day in parts of India, China, and Africa even though we now know that extensive contact

Microbiology of Infectious Disease. Sandy B. Primrose, Oxford University Press.
© Sandy B. Primrose (2022). DOI: 10.1093/oso/9780192863843.003.0013

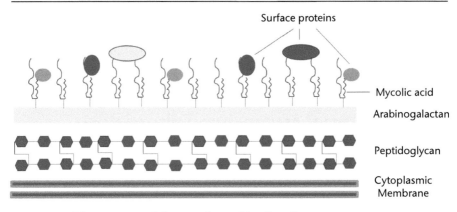

Surface proteins

Mycolic acid

Arabinogalactan

Peptidoglycan

Cytoplasmic Membrane

Figure 13.1. The structure of the mycobacterial cell wall.

is necessary to spread the infection. This poor infectivity of *Mycobacterium leprae* was demonstrated to the world in 1989 when Diana, Princess of Wales, visited a leprosy hospital in Indonesia and, against all advice, unreservedly touched a number of patients.

Infection with *Mycobacterium leprae* can lead to damage of the nerves, respiratory tract, skin, and eyes. This nerve damage can result in a decline in the ability to feel pain and leads to the loss of parts of a person's extremities, an obvious characteristic of people with leprosy, from repeated injuries or infection due to unnoticed wounds. An infected person may also experience muscle weakness and poor eyesight. The earliest written record of leprosy is from 600 BC and the earliest skeletal evidence dates from 300 BC. The oldest genomic evidence is the isolation of DNA from Asian samples dated to 80–240 AD. The disease is prevalent worldwide, with the exception of Europe, and there are ~ 200,000 new cases every year. The disease was widespread in mediaeval Europe but began declining in the sixteenth century and now is absent but the reason for the decline is not known. A severe form of leprosy, known as diffuse lepromatous leprosy (DLL), accounts for about 20% of cases in Mexico, Cuba, and Costa Rica, and is caused by *Mycobacterium lepromatosis*.

Mycobacterium leprae occasionally infects animals other than humans and the best known of these is the nine-banded armadillo found in the southern United States. DNA analysis of strains isolated from armadillos suggest that they are of human origin and that the animals acquired the disease from early American settlers. However, they are known to transmit leprosy back to humans today and thus make leprosy a zoonotic disease. Less well known is that *Mycobacterium leprae* has been found in red squirrels in England. In the mediaeval period there was a strong trade into eastern England of Baltic squirrels for the provision of fur. Also, at this time, squirrels were eaten as food and often kept as pets. The keeping of squirrels both domestically and in monasteries could have resulted in the transfer of leprosy bacteria from humans.

100

Tuberculosis (TB) is an infection caused by *Mycobacterium tuberculosis* or one of its close relatives such as *Mycobacterium bovis* or *Mycobacterium canettii*. TB generally affects the lungs, where it forms tubercles (Figure 13.2), but it can also affect other parts of the body. The majority of people who are

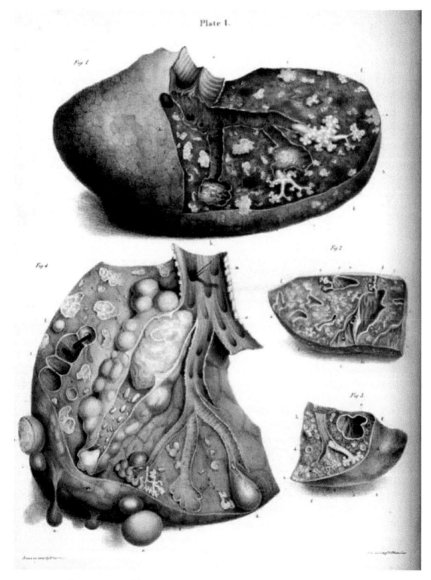

Figure 13.2. Illustration of lung tubercles made by Scottish doctor Sir Robert Carswell (1793–1857).

Source: University of Glasgow, UK via Wikimedia Commons (Public domain) <https://commons.wikimedia.org/wiki/File:Carswell-Tubercle.jpg>.

infected display no symptoms and are said to have latent TB. About 10% of latent infections progress to active disease which if left untreated, kills about half of those affected. The classic symptoms of active TB are a chronic cough with blood-containing mucus, fever, night sweats, and weight loss. It was this latter symptom that gave rise to the old name for the disease: consumption. In 15–20% of active cases, the infection spreads outside the lungs to sites such as the lymph glands in the neck (scrofula) and the spine (Potts' disease). Such extrapulmonary TB occurs more commonly in people with a weakened immune system, for example, people with HIV/AIDS.

TB has been a constant scourge throughout history and may have killed more people than any other infectious disease including 1 billion in the last 200 years. Hippocrates, in 400 BC, described a disease that he called phthisis where the symptoms were 'weakness of the lung' with fever and cough. It was the most common disease of the time and usually fatal. Around the same time, Aristotle provided the first written description of scrofula. Mycobacterial DNA and mycolic acids have been recovered from various archaeological sites that are ~ 10,000 years old. Initially it was assumed that domesticated animals were the source of TB that subsequently infected humans at these

---

Box 13.1 TUBERCULOSIS AND THE PASTEURIZATION OF MILK

Milk is an excellent medium for microbial growth and, when it is stored at ambient temperature, bacteria and other pathogens soon proliferate. Bacteria that can thrive in milk and milk products such as cheese include *Mycobacterium tuberculosis* (TB), *Brucella abortus* (brucellosis), *Corynebacterium diphtheriae*, *Listeria monocytogenes* (listeriosis and spontaneous abortion), and *Escherichia coli* O157:H7. The US Centers for Disease Control (CDC) says improperly handled raw milk is responsible for nearly three times more hospitalizations than any other food-borne disease source, making it one of the world's most dangerous food products. In 1887, James Law of Cornell University stated that bovine tuberculosis was a communicable disease and that humans could become infected from eating undercooked meat or fresh milk from tuberculous cattle. A few years later, in England, this view was endorsed by Medical Officers of Health and supported by various parliamentary enquiries.

In 1864, French chemist Louis Pasteur found out experimentally that the spoilage of wine could be prevented by heating it to about 50–60 °C (122–140 °F) for a short time to kill any microbes. The wine subsequently could be aged without sacrificing the final quality and this process was given the name 'pasteurization'. Pasteurization of milk was suggested by German chemist Franz von Soxhlet in 1886. The modern process of milk pasteurization involves heating it to 71.7°C for at least 15 seconds, and no more than 25 seconds, in a heat exchanger and then cooling it quickly to 3°C. Because of the nature of this heat treatment, it sometimes is referred to as the 'High Temperature Short Time' (HTST) process. All vegetative bacteria are killed by pasteurization, rendering the milk safe to drink, without any impact on taste or nutritional quality.

sites. However, all animal bones from these sites tested negative for *Mycobacterium* and genomic studies have shown that *Mycobacterium tuberculosis* is older than *Mycobacterium bovis*. These results suggest that animal domestication supported sizeable human populations and that the people inhabiting these ancient sites were infected with *Mycobacterium tuberculosis*. In more recent times, *Mycobacterium bovis* transmission from cattle to humans was common until such human infections were virtually eliminated by disease control in cattle herds and by routine pasteurization of cow's milk (Box 13.1).

The genomes of many different mycobacterial species have been sequenced and they show a remarkable variation in size (Table 13.1). This is strongly suggestive of horizontal gene transfer but the fact that the free-living species (*M. kansasii*, *M. marinarum*, and *M. smegmatis* in Table 13.1) have the largest genomes also points to genome reduction associated with pathogenesis. The most obvious example of genome reduction is provided by the organisms causing leprosy.

Genome analysis has allowed us to make comparisons between closely related organisms. For example, *Mycobacterium ulcerans* shared a common ancestor with *Mycobacterium marinarum* about 1 million years ago but the genome of the latter is one megabase larger. However, *Mycobacterium ulcerans* specifically carries a large (174,000 base pairs) plasmid that encodes mycolactone. This toxin causes tissue damage and inhibits the immune response that results in Buruli ulcers. Similarly, the two organisms that cause leprosy (*Mycobacterium leprae* and *Mycobacterium lepromatosis*) shared a common ancestor that underwent genome downsizing. However, since their separation ~ 14 million years ago, they have continued to lose genes but from different regions of the genome. Unfortunately, genome analysis has not given a clear explanation for the inability of the leprosy bacteria to be cultured in

Table 13.1. Genome data for some representative *Mycobacterium* species.

Organism	Genome size (Mb)	Number of genes
Mycobacterium avium	4.83	4,643
Mycobacterium bovis	4.3	3,951
Mycobacterium kansasii	6.42	6,007
Mycobacterium leprae	3.27	1,605
Mycobacterium lepromatosis	3.21	1,477
Mycobacterium marinarum	6.64	5,826
Mycobacterium smegmatis	6.46	6,591
Mycobacterium tuberculosis	4.4	4,553
Mycobacterium ulcerans	5.63	5,566

Note that the genome size may vary between different strains of the same species. The number of predicted genes depends on the software programme used.

the laboratory. Comparative genomics has shown that *Mycobacterium bovis* and *Mycobacterium tuberculosis* have greater than 99.95% homology, so both can cause TB in humans, but that the former has a large deletion that is not present in many strains of the latter.

Genome analysis also has provided an insight into the virulence mechanisms used by members of the *Mycobacterium tuberculosis* complex to survive and cause disease in their hosts. The most important virulence factors include the surface-exposed lipids in the mycobacterial outer membrane, type VII secretion systems (T7SS), and the Pro-Glu (PE)/ Pro-Pro-Glu (PPE) family proteins. All mycobacteria have a waxy outer surface resulting from the presence of mycolic acids. However, the pathogens also contain several unique lipids that allow them to survive *in vivo* by modulating the immune response. One way that they do this is to secrete the chemokine CCL2 which recruits monocytes to the site of infection. These monocytes provide the bacteria a safe haven from bactericidal macrophages.

The waxy outer layer of mycobacteria makes extracellular protein transport more difficult and so these bacteria utilize a specialized mechanism for protein transport known as a type VII secretion systems (T7SSs). In mycobacteria the T7SSs are known as ESX systems. In TB strains there are five ESX systems, and three (ESX-1, ESX-3, and ESX-5) have been implicated in virulence. ESX-1, for example, causes membrane disruptions in the host cell and allows *Mycobacterium tuberculosis* to escape destruction by phagocytes. Sequencing of the *Mycobacterium tuberculosis* genome revealed that 10% of its genes encode a unique family of proteins that have proline–glutamate and proline–proline–glutamate at their N-termini linked to a variable C-terminus. These PE and PPE proteins permit the bacterium to survive and spread by, among other things, increasing the levels of anti-inflammatory cytokines and inducing death of macrophages.

Mycobacterium kansasii is a soil-dwelling bacterium that is an occasional, opportunistic pathogen. Although its genome is much larger than that of *Mycobacterium tuberculosis*, the two genomes have a great deal of homology. Like the bacteria causing TB, *Mycobacterium kansasii* has five ESX systems and contains many PE/PPE proteins. Since it is found only rarely in patients, the ESX system and the PE/PPE protein family are not sufficient to explain the pathogenicity of the *Mycobacterium tuberculosis* complex. Comparative genomics has shown that *Mycobacterium tuberculosis* has fifty-five genes that are absent from *Mycobacterium kansasii*. These genes appear in clusters flanked by elements such as bacteriophage remnants and transposons that suggest that they were acquired by horizontal gene transfer. Some of these genes encode factors responsible for cell adhesion, arresting phagosome maturation, the production of compounds that function in oxidative stress resistance, and ones that modulate the host immune system.

Protection against TB can be induced by immunizing with either the BCG vaccine, *Mycobacterium bovis* BCG, or *Mycobacterium microti* that are members of the tightly knit *Mycobacterium tuberculosis* complex. The BCG strain was derived from a fully virulent isolate of *Mycobacterium bovis* by prolonged (eleven years!) serial passage that culminated in its attenuation, whereas *Mycobacterium microti* is naturally attenuated for humans even though it causes fulminant tuberculosis in voles and shrews. Over 3 billion individuals have been immunized with BCG without major side effects, while nearly 1 million humans have received the live vole bacillus vaccine. The stability of BCG and its lack of reversion to virulence suggested that an irreversible genetic event such as gene deletion could have contributed to the original attenuation process. Comparative genomics has confirmed this. It identified a series of chromosomal deletions common to both virulent and avirulent species but only a single locus, RD1, that has been deleted from *Mycobacterium bovis* BCG and *Mycobacterium microti*. The RD1 region encodes three proteins that are localized in the cell wall, and two of them, ESAT-6 and CFP-10, are also found in culture supernatants. These two proteins are yet another virulence mechanism of *Mycobacterium tuberculosis*: they prevent the human host from developing antibodies to the invading bacteria.

Vaccination is a cheap way of preventing a person from developing TB and certainly much cheaper than treating the disease. The usual treatment is isoniazid and rifampicin for six months and pyrazinamide and ethambutol for the first two months of the six-month treatment period. Unfortunately, antimicrobial resistance develops when treatment regimens are not adhered to or properly administered—a common problem in developing countries. Multidrug-resistant (MDR) tuberculosis entails resistance to isoniazid and rifampicin and extensively drug-resistant (XDR) TB includes additional resistance to any fluoroquinolone, and to any of the second-line injectable TB drugs (amikacin, capreomycin, and kanamycin). MDR-TB and XDR-TB are a growing problem worldwide and the only good news is that they is not caused by plasmid-borne antibiotic-resistance elements. Rather they are caused by mutations of genes encoding the targets for the antibiotics.

Because members of the *Mycobacterium tuberculosis* complex are so slow growing, it takes seven to eleven weeks to determine sensitivity to the first-line drugs and twelve to eighteen weeks to determine sensitivity to second-line drugs. The advent of genome sequencing has provided a method for greatly reducing the time for antibiotic sensitivity testing. Patient samples are incubated for two weeks in a proprietary Mycobacteria Growth Indicator Tube, as they would be for conventional antibiotic-sensitivity testing, and then DNA is extracted and sequenced. Sequencing and data analysis takes a maximum of two days and allows first-line and second-line antibiotic sensitivity to be determined with a high sensitivity and specificity in a total elapsed time of ~ two days—a time saving of five to sixteen weeks.

Mycobacterium tuberculosis has evolved to subvert, and even enrol for its own use, the mechanisms that the human immune system uses to clear bacterial infections. The host, in turn, is capable of limiting mycobacterial growth and inducing latency. However, after two decades of comparative genomics, we still do not have a detailed understanding of this bacterial–human host interaction. The same is true with leprosy where we know even less about the mechanisms of pathogenicity and what affects the balance between host immunity and bacterial evasion. Hopefully these issues will be resolved in the near future thereby permitting the development of new antibiotics to combat multi-drug resistance.

Key points

- The principal diseases caused by mycobacteria are tuberculosis and leprosy. Descriptions of both diseases can be found in pre-Christian Greek literature and DNA from both has been recovered from ancient skeletons.

- Leprosy is caused by *Mycobacterium leprae* and *Mycobacterium lepromatosis*. Leprosy still exists in many parts of the world but for unknown reasons, has died out in Europe. Tuberculosis is the best studied mycobacterial disease and is caused by members of the *Mycobacterium tuberculosis* complex (*M. tuberculosis, M. bovis, M. africanum*, and *M. canetti*).

- Genome analysis has shown that mycobacteria show a large variation in size and that the smallest genomes are found in the pathogenic species (genomic reduction). Genome analysis has not provided an explanation for the inability of *Mycobacterium leprae* to grow in the laboratory.

- Genomic analysis has revealed some of the key virulence factors of *Mycobacterium tuberculosis* and these include the type VII secretion system (known as ESX), the PE and PPE family of proteins, and mycolic acid. The latter helps to protect the bacterium from phagocytosis.

- Effective protection against tuberculosis is provided by vaccination with either *Mycobacterium microti* or the BCG strain of *Mycobacterium tuberculosis*. Both have a deletion of a genomic region encoding three cell wall proteins. Treatment of tuberculosis requires months of antibiotic therapy and multi-drug resistance is common. The use of DNA sequencing can greatly reduce the time to determine antibiotic sensitivity.

Suggested Reading

Behr M.A. (2015) Comparative genomics of mycobacteria: some answers, yet more questions. *Cold Spring Harbor Perspectives in Medicine* **5**(2), a021204. doi:10.1011/cshperspect.a021204

Bradley P., Gordon N.C., Walker T.M., Dunn L., Heys S., et al. (2015) Rapid antibiotic-resistance predictions from genome sequence data for *Staphylococcus aureus* and *Mycobacterium tuberculosis*. *Nature Communications* **6**, 10063. doi:10.1038/ncomms10063

Frith J. (2014) History of tuberculosis. Part 1—phthisis, consumption and the white plague. *Journal of Military and Veterans Health* **22**, 29–35.

Frith J. (2014) History of tuberculosis. Part 2—the sanatoria and the discoveries of the tubercle bacillus. *Journal of Military and Veterans Health* **22**, 36–41.

Saelens J.W., Viswanathan G., and Tobin D.M. (2019) Mycobacterial evolution intersects with host tolerance. *Frontiers in Immunology* **10**, 528. doi:10.3389/fimmun.2019.00528

Schuenemann V.J., Avanzi C., Krause-Kyora B., Seitz A., Herbig A., et al. (2018) Ancient genomes reveal a high diversity of *Mycobacterium leprae* in mediaeval Europe. *PLOS Pathogens* **14**(5), e1006997. doi:10.1371/journal.ppat.1006997

Singh P., Benjak A., Schuenemann V.J., Herbig A., Avanzi C., et al. (2015) Insight into the evolution and origin of leprosy bacilli from the genome sequence of *Mycobacterium lepromatosis*. *Proceedings of the National Academy of Sciences* **112**, 4459–64.

14

Plasmids and Pathogenicity: The *Bacillus cereus* Complex

Two genera of bacteria, *Bacillus* and *Clostridium*, have a common structural feature that makes them stand out from most other bacteria: they produce heat-resistant endospores that also give protection from desiccation and ultraviolet (UV) exposure. The two genera are distinguished by their ability to grow in the presence of oxygen. *Bacillus* species can grow in the presence of oxygen whereas *Clostridium* species cannot. Classically, microbiologists recognized another difference between the two genera. Whereas most *Clostridium* species are human pathogens, almost all *Bacillus* species are non-pathogenic. The one exception to this rule was *Bacillus anthracis*, the causative organism of anthrax.

Bacillus anthracis causes different forms of anthrax depending on how it enters the body. Cutaneous anthrax (Figure 14.1) occurs when anthrax spores get into the skin, usually through a cut or scrape. This can happen when a person handles infected animals or contaminated animal products like wool, hides, or hair. Cutaneous anthrax is the most common form of anthrax infection, and the least dangerous. Inhalation anthrax occurs when a person breathes in anthrax spores. This is the deadliest form of anthrax and even with aggressive antibiotic therapy patients only have a 50% chance of survival. In the United Kingdom, inhalation anthrax was known as woolsorter's disease, a much-feared condition that was very common in the late nineteenth century among people who sorted wool. The study of the disease identified *Bacillus anthracis* as the cause and led to the development of ventilation systems to take dust away from workers. Gastrointestinal anthrax is very rare as it occurs when a person eats raw or undercooked meat from an animal infected with anthrax.

Microbiology of Infectious Disease. Sandy B. Primrose, Oxford University Press.
© Sandy B. Primrose (2022). DOI: 10.1093/oso/9780192863843.003.0014

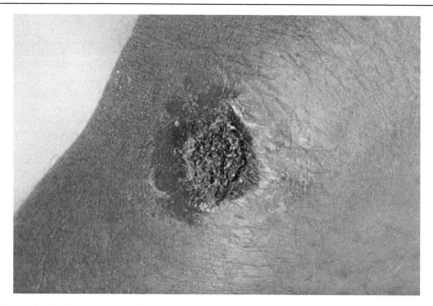

Figure 14.1. Cutaneous anthrax.
Source: Centers for Disease Control and Prevention, National Center for Emerging and Zoonotic Infectious Diseases (NCEZID) (Public domain) <https://www.cdc.gov/anthrax/lab-testing/recommended-specimens/index.html>.

Anthrax is a zoonosis and both cutaneous and inhalation anthrax now are extremely rare in developed countries as farm animals in these countries are free of the disease. The last recorded case of inhalation anthrax in the United Kingdom was in 2008 and occurred in a person making bongo drums using imported animal hides. After fixing the hides to the drum chassis, a metal scraper was used to remove the animal hair and this released an aerosol of anthrax spores. A new form of anthrax, injectional anthrax, has been discovered recently. It first was recorded in 2009 in Scotland and led to fourteen fatalities. Since then, other cases have been reported in various European countries. In each case, the disease was linked to contamination of illegal heroin, possibly from Afghanistan.

Despite its rarity, anthrax is much feared in the developed world because of its potential as a biological weapon (Box 14.1). *Bacillus anthracis* spores can be weaponized easily and, even with antibiotic therapy, large numbers of victims would die from inhalation anthrax. The Iraqi dictator Saddam Hussein had an extensive bioweapons programme that included anthrax. During the First Gulf War (1990–1991), allied soldiers had to don very cumbersome protective clothing after every enemy rocket strike in case anthrax had been released. The biggest downside of anthrax as a bioweapon is that its use leaves land uninhabitable. During the Second World War, the British government tested the efficacy of anthrax as a bioweapon by releasing spores

---

Box 14.1 ANTHRAX AS A BIOWEAPON

The first use of anthrax as a bioweapon is believed to be by Germany during the First World War. There is evidence that the German army used *Bacillus anthracis* to secretly infect livestock and animal feed traded to the Allied Nations by neutral partners such as Argentina. In 1932, during their occupation of Manchuria, the Japanese infected prisoners with anthrax and other deadly diseases. It was later discovered that they attacked at least eleven Chinese cities with anthrax by spraying it directly onto homes from aircraft.

In 1979, sixty-four people in the Russian town of Sverdlovsk died after an unintended release of anthrax spores from a biological weapons facility. The authorities tried to claim that the victims had died from gastrointestinal anthrax but Western governments were sceptical. In 1992, then-president of Russia, Boris Yeltsin, admitted that the outbreak was exactly what Western analysts had determined.

In 1993, the Japanese Aum Shinrikyo cult sprayed a liquid suspension of *Bacillus anthracis* from their headquarters building in Kameido, near Tokyo, Japan. This release went unnoticed. The cult's later (1995) sarin gas attack of a Tokyo subway attracted worldwide attention. It was only with testimony of cult members and a retrospective investigation that the 1993 incident was recognized as an anthrax release.

In 2001, letters filled with a white powder containing anthrax spores were mailed to two US Senators' offices and news media agencies along the East Coast of the United States. The powder form allowed the anthrax to float in the air and for it to be breathed in resulting in inhalation anthrax and five people died. The powder from these letters contaminated the postal facilities they were processed through as well as the buildings where they were opened. Genetic typing of the anthrax strain involved identified from where it had come originally but the perpetrator was not identified.

---

from bombs on the uninhabited Gruinard Island off the north-west coast of Scotland. The effectiveness of the weapons was judged by tethering sheep downwind of the detonation point. Fifty years later, the island still was contaminated attesting to the longevity of bacterial spores. Indeed, the island was only freed of the organism after 10 acres of it were decontaminated with formaldehyde.

The lethality of *Bacillus anthracis* is principally due to two virulence factors: a capsule (Figure 14.2) composed of polyglutamic acid which prevents the bacteria from being phagocytosed, and anthrax toxin. The toxin is a three-protein exotoxin secreted by the bacteria and composed of a cell-binding protein, known as protective antigen (PA), and two enzyme components, called oedema factor (EF) and lethal factor (LF). PA plus LF produces lethal toxin, and PA plus EF produces oedema toxin (Figure 14.3). These toxins cause cell death and tissue swelling (oedema), respectively. Interestingly, the toxin was first purified by Professor Harry Smith working on an open bench (!) because at the time there were no negative pressure containment hoods. The Sterne strain that is used to make anthrax vaccine can still produce the oedema factor and the lethal factor but it lacks the capsule, which indicates the importance of the capsule for virulence.

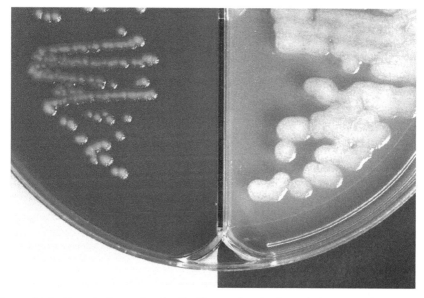

Figure 14.2. Capsule formation by *Bacillus anthracis*. Capsule-free growth on blood agar (left) and profuse capsule formation on bicarbonate agar (right).

Source: Pixnio/Dr James Feely, USCDCP, reproduced under Creative Commons CCO licence <https://pixnio.com/science/microscopy-images/anthrax-bacillus-anthracis/bacillus-anthracis-positive-encapsulation-test-rough-colonies-on-blood-agar-and-smooth-colonies-on-bicarbonate-agar>.

It has long been known that, in terms of its biochemical properties, *Bacillus anthracis* is very closely related to soil-dwelling *Bacillus cereus*. Initially, it was considered that the latter organism was non-pathogenic but towards the end of the twentieth century it was realized that it could cause food poisoning. Indeed, two types of food poisoning are recognized: a diarrheal form associated with a wide range of foods and an emetic form associated with cooked rice. The two types of disease are caused by different toxins. The emetic toxin is a cyclic peptide called cereulide whose mode of action is believed to be caused by binding to, and activation of, 5-HT3 receptors, leading to increased afferent vagus nerve stimulation. Diarrhoeal disease is caused by a combination of the three toxins: haemolysin BL (Hbl), nonhaemolytic enterotoxin (Nhe), and cytotoxin K (CytK). These enterotoxins are all produced in the small intestine of the host,

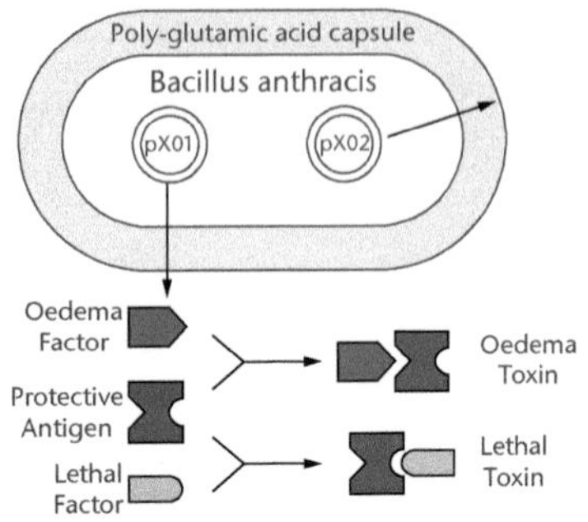

Figure 14.3. The virulence factors of *Bacillus anthracis*. The plasmid pX01 encodes the various toxin components and plasmid pX02 encodes the poly-glutamic acid capsule.

Another organism that is closely related to *Bacillus cereus* and *Bacillus anthracis* is *Bacillus thuringiensis*. Upon sporulation, this bacterium forms crystals of proteinaceous insecticidal toxins that contain two types of proteins: Cry (crystal) and Cyt (cytolytic) proteins. There are different variants of the toxin and each variant has a spectrum of insects such as moths, flies, mosquitos, bees, etc., against which they are toxic. These toxins have been used for agricultural pest control (Box 14.2). Vegetative cells of *Bacillus thuringiensis* also can produce two toxins: vegetative insecticidal protein (VIP) and secreted insecticidal protein (SIP).

Box 14.2 AGRICULTURAL PEST CONTROL WITH *BACILLUS THURINGIENSIS*

For over sixty years, the spores and crystalline insecticidal proteins produced by *Bacillus thuringiensis* have been produced commercially and used as spray insecticides to protect specific crops. Approximately 180 such products have been registered in the United States and 120 in the European Union. Each year, China uses > 10,000 tons (!) of the bacteria in its agriculture. The extensive use of *Bt* microbial pesticides worldwide is likely to be due to their specificity against a limited number of target insect species that greatly limits the potential for impacts to beneficial and non-target organisms.

Because of the efficacy, safety to humans and non-target organisms, and the favourable environmental persistence profile of *Bt* microbial formulations, the active proteins in these formulations have been isolated and optimized for expression in plants to make genetically modified (GM) crops (Figure 14.4) that are resistant to target insects. In 1996, GM maize producing the *cry* protein to kill the European corn borer received regulatory approval in the United States. Later, additional toxin genes were introduced that killed corn rootworm larvae. Corn modified to produce VIP was approved in 2010. The other major crop engineered to produce *Bt* toxins is cotton.

Figure 14.4. Bt toxins present in transgenic peanut leaves (right) protect them from extensive damage caused to unprotected peanut leaves by lesser cornstalk borer larvae (left).

Source: Agricultural Research Service, United States Department of Agriculture, with the ID K8664-1 (Public domain) <https://commons.wikimedia.org/wiki/File:Bt_plants.png>.

The genomes of a large number of strains of *Bacillus anthracis*, *Bacillus cereus*, and *Bacillus thuringiensis* have been sequenced and two things now are clear. First, they are distinct organisms but they are more closely related to one another than to any other of the more than 100 *Bacillus* species. For this reason, they often are referred to as the *Bacillus cereus* group. Second, their different pathogenicity profiles are due to the presence of different plasmids carrying genes that encode various pathogenicity factors. Classical strains of *Bacillus anthracis* carry two plasmids essential for full virulence, pOX1 encoding anthrax toxin and the other, pOX2 encoding the polyglutamic acid capsule (Figure 14.3). In *Bacillus thuringiensis*, the genes responsible for the production of the insecticidal toxin (Cry protein) also lie on plasmids. Emetic strains of *Bacillus cereus* carry a plasmid that specifies the toxin cereulide. However, since 2004 there have been reports of atypical *Bacillus cereus* strains causing anthrax-like disease in humans. These strains have chromosomal genomes characteristic of *Bacillus cereus* but carry plasmids related to pOX1 and pOX2. The only toxins not encoded on plasmids are the Hbl, Nhe, and CykK toxins that cause diarrhoeal disease. Interestingly, the genes for these toxins are found in *Bacillus anthracis* and *Bacillus thuringiensis*, as well as *Bacillus cereus*, but are inactivated.

The genomic structure of the *Bacillus cereus* group suggests that diarrhoeal strains of *Bacillus cereus* are the parent and that *Bacillus anthracis* and *Bacillus thuringiensis* were generated by the acquisition of plasmid-borne toxin genes. This begs the question: from where did these toxin genes come? Anthrax toxin is known as a binary bacterial toxin and other binary toxins occur in *Clostridium* species. However, these toxins do not have much similarity. Again, bacteria that produce toxins similar to the Cry toxins of *Bacillus thuringiensis* are unknown. The only toxin in the *Bacillus cereus* group that has a homologue is cereulide. This has a similar structure to valinomycin produced by *Streptomyces* species and, like it, has a strong affinity for potassium ions.

Another enigmatic aspect of the *Bacillus cereus* group is the numbers and sizes of the plasmids that they carry. Diarrhoeal strains of *Bacillus cereus* are largely free of plasmids and so the chromosome must carry all the genes required for growth outside of animals. Maintaining plasmids in cells in the absence of selective pressure has an energetic cost and the larger the plasmid the greater the cost. Despite this, many strains of the *Bacillus cereus* group contain multiple plasmids—and large ones at that. For example, for *Bacillus anthracis* to be virulent it needs to carry plasmids pOX1 and pOX2 whose sizes are ~ 180,000 base pairs and 96,000 base pairs respectively. Between them, these two plasmids encode over sixty genes—but what do these genes do? Many of them encode proteins with no homology to known proteins. Also, why not have the two key pathogenicity determinants (capsule formation and anthrax toxin) encoded on a single plasmid? Although anthrax is a zoonosis, it does not appear to be very common, even in wild animals. So, what is the selective pressure?

The plasmids encoding the *Bacillus thuringiensis* toxins also are large, ranging in size from 150,000 to 580,000 base pairs. Given that *Bacillus thuringiensis* is widespread in the soil and infects many different insects it is likely that there is selective pressure for plasmid retention. Up to fifty genes on the *Bacillus thuringiensis* plasmids have homology with genes on the pOX1 plasmid of *Bacillus anthracis* implying that they have a role in pathogenesis—but what role? Some *Bacillus thuringiensis* strains carry a plasmid with as many as five Cry protein genes and three Cyt protein genes. Other strains carry multiple plasmids that each carry a *cry* gene. What is the benefit of these different genomic structures? A comparative genomic analysis of *Bacillus thuringiensis* strains showed that highly toxic strains contained significantly more insecticidal toxicity-related genes thereby providing strategies for infection, immune evasion, and insect cadaver utilization. This fact, plus the multiple *cry* genes may account for plethora of strains with different insect specificities. Clearly, we have much to learn about the natural history of the *Bacillus cereus* group of pathogens, and *Bacillus anthracis* and *Bacillus thuringiensis* in particular.

Key points

- *Bacillus cereus, Bacillus anthracis,* and *Bacillus thuringiensis* comprise the *Bacillus cereus* complex. They are more related to one another than to all other *Bacillus* species.

- *Bacillus anthracis* causes anthrax and there are different forms of the disease: cutaneous, inhalation, and injection. If used as a bioweapon, *Bacillus anthracis* will cause a fatal inhalation anthrax. The key virulence factors are the plasmid-encoded capsule and components of the oedema toxin and the lethal toxin.

- *Bacillus cereus* can cause two types of food poisoning: diarrhoeal disease and emetic disease. Emetic disease is caused by a plasmid-encoded toxin cereulide. Diarrhoeal disease is caused by three chromosomally encoded toxins.

- *Bacillus thuringiensis* strains produce a number of plasmid-encoded toxins that are insecticidal. Different toxins are lethal for different insects.

- The parent of the *Bacillus cereus* complex most likely was a diarrhoeal strain of *Bacillus cereus* because it has no plasmid-borne toxins. This strain then acquired different plasmids to generate the different pathogens in the complex as we know it today.

Suggested Reading

Baldwin V.M. (2020) You can't *B. cereus*—a review of *Bacillus cereus* strains that cause anthrax-like disease. *Frontiers in Microbiology* **11**, 1731. doi:10.3389/fmicb.2020.01731

Ehling-Schulz M., Lereclus D., and Koehler T.M. (2019) The *Bacillus cereus* group: *Bacillus* species with pathogenic potential. *Microbiology Spectrum* **7**, 3. doi:10.1128/microbiolspec.GPP3-0032-2018

Koch M.S., Ward J.M., Levine S.L., Baum J.A., Vicini J.L., and Hammond B.G. (2015) The food and environmental safety of *Bt* crops. *Frontiers in Plant Science* **6**, 283. doi:10.3389/fpls.2015.00283

Manchee R.J., Broster M.G., Stagg A.J., and Hibbs S.E. (1994) Formaldehyde solution effectively inactivates spores of *Bacillus anthracis* on the Scottish island of Gruinard. *Applied and Environmental Microbiology* **60**, 4167–71.

Pilo P. and Frey J (2018) Pathogenicity, population genetics and dissemination of *Bacillus anthracis*. *Infection, Genetics and Evolution* **64**, 115–25. doi:10.1016/j.meegid.2018.06.024

Zhu L., Peng D., Wang Y., Ye W., et al. (2015) Genomic and transcriptomic insights into the efficiency of *Bacillus thuringiensis*. *Scientific Reports* **5**, 1–11.

15

Tracking the Origins of *Clostridium difficile* Infections

The organism known as *Clostridium difficile* (Box 15.1) was first isolated in 1935 and it was thought to be a harmless occupant of the mammalian gut. However, in 1978 it was recognized as a pathogen when it was found to cause antibiotic-associated pseudomembranous colitis, a swelling and inflammation of the large intestine (colon). Since then, the bacterium has become the most common cause of antibiotic-associated diarrhoea and healthcare infection in the developed world. The disease it causes is known as *Clostridium difficile*-associated disease (CDAD). Pathogenesis is associated with the production of two toxins, TcdA and TcdB, that inhibit the polymerization of actin in epithelial cells resulting in the death of these cells (Figure 15.1). The tight junctions between intestinal epithelial cells then break down leading to fluid loss (diarrhoea). In mild forms of CDAD the patient produces watery diarrhoea at least three times per day and may have abdominal pain. In the most severe forms of CDAD, the patient also has fever, a high white blood cell count, intestinal bleeding, and dehydration and low blood pressure because of excessive fluid loss. In the most critical form of CDAD, the colon becomes dilated (toxic megacolon) and is at risk of perforation.

Box 15.1 A TAXONOMIC NOTE

Classically trained microbiologists were taught that certain bacteria could form spores that were resistant to heat, desiccation, and many chemical agents. Furthermore, these spore-forming bacteria fell into two genera: the aerobic spore-formers (*Bacillus* species) and the anaerobic spore-formers (*Clostridium* species). Today there are over eighty genera of bacteria that produce spores, although not all of them necessarily display the characteristic heat resistance. To further complicate matters, DNA typing methods and genome sequencing have shown that the organism known as *Clostridium difficile* is a distant relative of most other *Clostridium* species, particularly the human

continued

Microbiology of Infectious Disease. Sandy B. Primrose, Oxford University Press.
© Sandy B. Primrose (2022). DOI: 10.1093/oso/9780192863843.003.0015

Box 15.1 *Continued*

pathogens such as *Clostridium tetani*, *Clostridium botulinum*, and *Clostridium perfringens*. As a consequence, it now is common practice to refer to *Clostridium difficile* as *Clostridioides difficile* but the original nomenclature is used in this text.

A person normally gets infected with *Clostridium difficile* when they ingest spores. These spores pass though the gastrointestinal tract unchanged until they reach the colon where the presence of bile induces them to germinate and become vegetative cells. The vegetative cells then produce the TcdA and TcdB toxins and cause CDAD. Many of the vegetative cells then sporulate, although why they do this is not known, and pass out of the body in faeces or diarrhoeal discharges (Figure 15.1). *Clostridium difficile* seldom

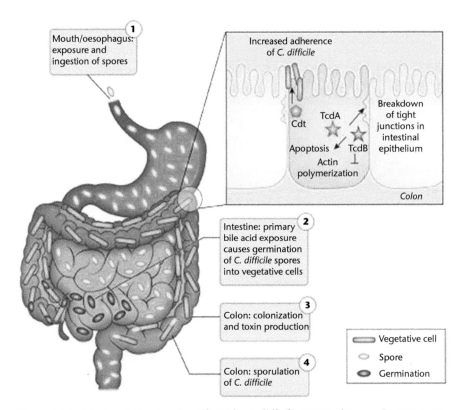

Figure 15.1. Mechanism whereby *Clostridium difficile* causes disease. Figure reproduced by permission of Elsevier from Sandhu B.K. and McBride S.M. (2018) *Clostridioides difficile. Trends in Microbiology* 26(12), 1049–50.

causes problems in people with a normal gut flora and many people normally harbour the bacterium in their gut, albeit at a low level. The problem with CDAD arises when patients are treated with antibiotics that kill off the normal gut flora, particularly clindamycin, broad spectrum penicillins (see p85), cephalosporins, and fluoroquinolones (see p43). Patients with serious illnesses and prolonged hospitalization are particularly at risk, as are people above sixty-five years of age.

Following the recognition of *Clostridium difficile* as the cause of antibiotic-associated diarrhoea and colitis, standard hospital practice was to stop administering the antibiotic causing the problem (usually cephalosporins) and to treat the patient with either vancomycin or metronidazole. In 2003 a major outbreak occurred in Canada where the symptoms of infection were much more severe and were refractory to standard treatment. This outbreak was followed by ones in the United States and several in the United Kingdom with high numbers of deaths. In each case the outbreak was caused by a hypervirulent strain of *Clostridium difficile* that belonged to a ribotype (see Box 15.2) designated as 027. So, what led to the predominance of this ribotype? One interesting proposal is that its emergence is linked to the widespread introduction of the sugar trehalose by the food industry around 2001. Strains belonging to ribotype 027 can grow on the low levels of trehalose typically found in the gut whereas strains belonging to many other ribotypes cannot. However, this explanation has been disputed. Another possible explanation is that strains belonging to ribotype 027 are resistant to fluoroquinolones, antibiotics (see p43) whose use increased greatly in the 1990s.

Box 15.2 PCR RIBOTYPING OF *CLOSTRIDIUM DIFFICILE*

Ribotyping is the standard method for identifying different strains of *Clostridium difficile*. Ribosomal DNA (rDNA) consists of the genes for the 16S and 23S ribosomal RNA genes separated by a spacer region. There are multiple copies of rDNA and the size of the spacer region varies between each copy. PCR ribotyping of *Clostridium difficile* involves the amplification of rDNA across the spacer regions using primers that are complementary to the 3' end of the 16S rRNA gene and the 5' end of the 23S rRNA gene. Electrophoretic resolution of the DNA amplicons generates profiles which are interpreted as a fingerprint or *ribotype* that can be used to distinguish *Clostridium difficile* strains from one another.

continued

Box 15.2 *Continued*

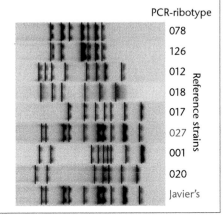

Figure 15.2. A gel showing PCR products of various *Clostridium difficile* strains. The patient's sample (Javier's) is shown at the bottom and it matches ribotype 27 in the reference set. <https://ecampusontario.pressbooks.pub/microbio/chapter/visualizing-and-characterizing-dna-rna-and-protein/> Photo reproduced courtesy of Dr Wendy Keenleyside.

PCR-ribotype: 078, 126, 012, 018, 017, 027, 001, 020 (Reference strains), Javier's

The genome of *Clostridium difficile* is ~ 4.3 Mb in size but it demonstrates extreme plasticity and only about 16% of the genome is shared between different strains. All toxigenic strains have a 19.6 kilobase region of the chromosome called the Pacloc locus that contains the genes for the TcdA and TcdB toxins. However, no difference has been seen in this locus between pre-and post-epidemic isolates of ribotype 027. There also is a separate locus (*agr*) encoding a quorum sensing system (see p82) that controls toxin production. That is, toxin is only produced when the population of *Clostridium difficile* exceeds a certain level as happens when the normal gut flora has been eliminated by antibiotics. Significant differences have been seen in the *agr* loci between hypervirulent and non-hypervirulent strains. When genome sequencing was applied to multiple isolates of ribotype 027 it showed that its core genome exhibited very little genetic diversity, which is in marked contrast to the species as a whole. This low level of diversity is consistent with recent emergence. All the ribotype 027 strains had acquired high-level resistance to fluoroquinolone antibiotics but there were two lineages: FQR1 and FQR2. The only differences between these lineages were the mutation in the *gyrA* gene causing the fluoroquinolone resistance and the presence of a transposon (Tn6105) in FQR2. Both FQR1 and FQR2 arose in North America but they had different patterns of global spread. FQR1 caused sporadic cases in Switzerland and South Korea but FQR2 spread to Europe on at least four occasions.

The biggest risk of developing a *Clostridium difficile* infection comes from a current or recent stay in a hospital. Hospitals can minimize the incidence of such infections by careful antibiotic stewardship. For example, infections with fluoroquinolone-resistant strains of *Clostridium difficile* declined in the

United Kingdom following the reduction in use of these antibiotics. However, infections will always occur. When they do, the patient will excrete over 1 million spores per gram of faeces making environmental contamination highly likely, especially when there is diarrhoea. Given the considerable time during which patients may shed spores, prompt isolation of symptomatic cases and adequate environmental decontamination are key recommendations for preventing onward transmission. Curiously, using whole genome sequencing to identify individual strains, it has been shown that less than 40% of hospital infections with *Clostridium difficile* are acquired from other symptomatic patients. This suggests that the major source of infection is asymptomatic patients who carry *Clostridium difficile* in their large intestine. If this is the case then asymptomatic patients need to be identified on admission to hospital and isolated.

A major study of *Clostridium difficile* diversity in Europe was done using hospital inpatient diarrhoeal samples collected on two specific days from 482 hospitals. All the samples were ribotyped and those belonging to the ten most prevalent ribotypes had their whole genomes sequenced. This study revealed two distinct patterns of *Clostridium difficile* spread. Certain ribotypes, such as 027 and 001, were associated with healthcare facilities and circulated locally. Other ribotypes, such as 014 and 078, showed no evidence of local clustering: rather, they were disseminated widely across Europe and were particularly associated with pig farming. Ribotype 078 is of particular concern because it causes a very severe disease with high mortality, it is community-associated, and causes more infections in younger age groups.

Ribotype 078 has been found in sick and healthy animals, particularly pigs, bird droppings, vermin, and the farm environment, as well as a variety of retail meat products such as pork and beef. These natural reservoirs suggest that humans become colonized with ribotype 078 via the food chain and/or the environment. This idea is supported by analysis of its antibiotic sensitivity. Most strains are sensitive to fluoroquinolones and clindamycin, the use of which has driven the rise of *Clostridium difficile* infections in hospitals in Europe and the United States respectively. By contrast, most strains of ribotype 078 were resistant to tetracyclines, antibiotics that inhibit bacterial protein synthesis by binding to the bacterial ribosome. There are a number of different modes of tetracycline resistance but in ribotype 078 it is due to the presence of the *tetM* gene which encodes a ribosome protection protein. The clinical use of tetracyclines is restricted mostly to young persons with acne and sexually transmitted chlamydial disease and is unlikely to have driven the selection of ribotype 078. However, tetracyclines are the most widely used antibiotics for the treatment and prevention of infections in animals and also are used for growth promotion and so are a likely source of selective pressure.

Like ribotype 078, many ribotype 014 isolates are resistant to tetracycline as a result of acquisition of the transposon-carried *tetM* gene. In Australia, ribotype 014 is common in farm animals, particularly pigs, and is associated with many cases of community-acquired *Clostridium difficile* infections. However, most patients in Australia who get infected with ribotype 014 have never been near a pig, or a farm for that matter, so from where did they acquire it? *Clostridium difficile* spores are abundant in treated biosolids, effluent, and piggery wastewater. These by-products of the pig industry are subsequently recycled to pasture and agriculture for composting and direct irrigation/fertilization of crops and lawns. Not surprisingly, *Clostridium difficile* has been recovered from plant composts sold by retailers as well as from root vegetables from mainstream and organic markets. Eliminating this source of infection will be extremely difficult.

Many of the problems associated with *Clostridium difficile*, be it recurrence in treated patients or persistence in the environment, are associated with the production of spores. In the case of recurrence in patients, a key contributing factor is the persistence of spores in the intestinal lumen following treatment. These spores are unaffected by standard antimicrobial therapy, and provide a key reservoir for recurrent disease, allowing re-establishment of a vegetative *Clostridium difficile* population following successful initial treatment. Fidaxomicin is a recently introduced antibiotic which may provide a solution to this problem. First, it selectively eradicates *Clostridium difficile* with relatively little disruption to the multiple species of other bacteria that make up the normal, healthy intestinal microbiota. This maintenance of normal physiological conditions in the colon may reduce the probability of recurrence of *Clostridium difficile* infection. Second, and more important, fidaxomicin persists on spores, whereas vancomycin and metronidazole do not. This persistence prevents subsequent spore outgrowth and toxin production. Whilst of great clinical value, fidaxomicin cannot be used to eliminate agricultural sources of infection.

Key points

- *Clostridium difficile* is a major cause of hospital-acquired infections and rates of infection are directly linked to the use of broad-spectrum antibiotics. The bacterium can be present asymptomatically in the gut and only becomes a problem when the normal gut flora is eliminated by antibiotics.

- The bacterium produces two toxins, TcdA and TcdB, that cause death of intestinal epithelial cells and diarrhoea. When spores of *Clostridium difficile* are ingested, bile induces their germination to vegetative cells that produce the toxins. The cells then sporulate and are excreted in large numbers with the diarrhoea.

- There are many different ribotypes of *Clostridium difficile* and certain ones appear to be hypervirulent although the reason why is not known. Certain of these hypervirulent ribotypes are hospital-acquired and tend to be resistant to fluoroquinolones that are widely used clinically. Other hypervirulent ribotypes are tetracycline-resistant and these are linked with farm animals, especially where tetracycline has been used as an animal supplement.

- Fidaxomicin is a new antibiotic that coats spores in the gut and prevents spore outgrowth and toxin production.

Suggested Reading

Chilton C.H., Crowther G.S., Ashwin H., Longshaw C.M., and Wilcox M.H. (2016). Association of fidaxomicin with *C. difficile* spores: effects of persistence on subsequent spore recovery, outgrowth and toxin production. *PLoS One* **11**(8), e0161200. doi: 10.1371/journal.pone.0161200

Eyre D.W., Davies K.A., Davis G., Fawley W.N., Dingle K.E., et al. (2018) Two distinct patterns of *Clostridium difficile* diversity across Europe indicating contrasting routes of spread. *Clinical Infectious Diseases* **67**(7), 1035–44. doi:10.1093/cid/ciy252

He M., Miyajima F., Roberts P., Ellison L., Pickard D., et al. (2013) Emergence and spread of epidemic healthcare-associated *Clostridium difficile*. *Nature Genetics* **45**, 109–13.

Knetsch C.W., Kumar N., Forster S.C., Connor T.R., Browne H.P., et al. (2018) Zoonotic transfer of *Clostridium difficile* harboring antimicrobial resistance between farm animals and humans. *Journal of Clinical Microbiology* **56**(3), e01384–17. doi:10.1128/JCM.01384-17

Knight D.R., Elliot B., Chang B.J., Perkins T.T., and Riley T.V. (2015) Diversity and evolution in the genome of *Clostridium difficile*. *Clinical Microbiology Reviews* **28**, 721–41.

Knight D.R., Squire M.M., Collins D.A., and Riley T.M. (2017). Genome analysis of *Clostridium difficile* PCR ribotype 014 lineage in Australian pigs and humans reveals a diverse genetic repertoire and signatures of long-range interspecies transmission. *Frontiers in Microbiology* **28**(3): 721–41. doi:10.3389/fmicb.2016.02138

O'Hagan J.J. and McDonald L.C. (2018) The challenges of tracking *Clostridium difficile* to its source in hospitalized patients. *Clinical Infectious Diseases* **68**(2), 210–12. doi:10.1093/cid/ciy461

16

Tracking Horizontal Gene Transfer: *Staphylococcus aureus*

Staphylococcus aureus is part of the normal human microbial flora and colonizes diverse niches including the upper respiratory tract, the nasal passages, and skin. If the environmental conditions are favourable, it will cause disease, and it is the leading cause of skin and soft tissue infections and a common cause of osteomyelitis, endocarditis, and pneumonia. It is a major cause of hospital-acquired (nosocomial) infections, particularly of wounds post-operatively. Until the late 1940s, when penicillin became widely available, an infection with *Staphylococcus aureus* was much feared because it had a high probability of causing death. Unfortunately, penicillin resistance soon became widespread due to the bacterium acquiring a gene encoding penicillinase (β-lactamase), an enzyme that destroys the β-lactam ring of the penicillin molecule. Penicillinase-resistant β-lactam antibiotics (Box 16.1) such as methicillin were developed in the late 1950s but *Staphylococcus aureus* soon acquired resistance to these as well. Today, methicillin-resistant *Staphylococcus aureus* (MRSA) is of major concern, particularly in hospital settings.

Box 16.1 PENICILLINS AND THEIR MODE OF ACTION

Penicillins are a group of antibiotics originally obtained from culture broths of the fungi *Penicillium notatum* and *Penicillium chrysogenum*. Most penicillins in clinical use are chemically synthesized from naturally produced penicillins. The fungi produce a number of different penicillins but only two have been used clinically: penicillin G (intravenous use) and penicillin V (given by mouth). The penicillins have a characteristic structure as shown in Figure 16.1) where R equals a variable side chain. In penicillin G, R is a benzyl residue and in penicillin V, it is a phenoxymethyl residue. The R group

Microbiology of Infectious Disease. Sandy B. Primrose, Oxford University Press.
© Sandy B. Primrose (2022). DOI: 10.1093/oso/9780192863843.003.0016

can be cleaved from natural penicillins to produce 6-aminopenicilloic acid and a new R group added chemically to generate semi-synthetic penicillins. The semi-synthetic penicillins that are resistant to β-lactamase are cloxacillin, dicloxacillin, fluoxacillin, methicillin (also known as meticillin), and oxacillin.

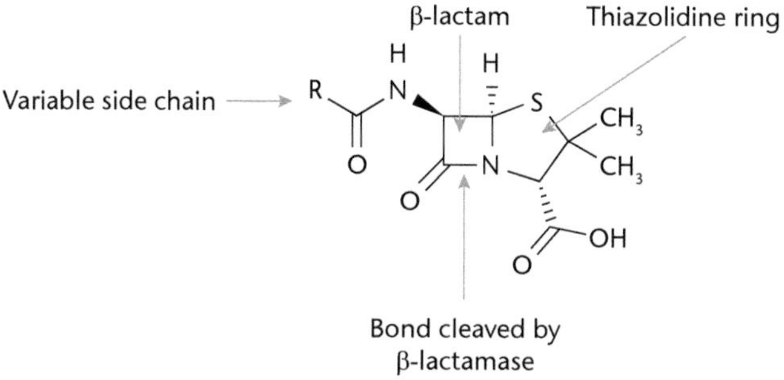

Figure 16.1. Key features of a penicillin molecule.

Penicillins kill bacteria by inhibiting the synthesis of the peptidoglycan layer of bacterial cell walls. The peptidoglycan layer is important for cell wall structural integrity, especially in *Staphylococcus aureus*, and inhibition of its synthesis results in cell lysis. The final step in the synthesis of the peptidoglycan is facilitated by DD-transpeptidases, also known as penicillin-binding proteins (PBPs). The normal substrate for these enzymes is D-alanyl-D-alanine but penicillins have a structural similarity and their β-lactam nucleus can bind irreversibly to the active site and prevent normal peptidoglycan synthesis.

Staphylococcus aureus produces a wide range of virulence factors (Figure 16.2) but not all isolates have all factors. One of these factors is the pigment staphyloxanthin which gives bacterial colonies a golden colour that is reflected in the bacterium's specific epithet 'aureus'. This pigment is an antioxidant and it helps the bacterium evade the reactive oxygen species that the host immune system uses to kill pathogens. Other virulence factors include enzymes such as coagulase and staphylokinase. Coagulase clots plasma which then coats the bacterial cell and prevents phagocytosis. Staphylokinase dissolves fibrin, which normally forms blood clots, and this will aid in spread of the bacterium. To resist phagocytic clearance, *Staphylococcus aureus* expresses a polysaccharide capsule which effectively masks the bacterial surface and surface-associated proteins from recognition by phagocytic cells. This is one of the virulence mechanisms used by strains that cause mastitis in cows.

Secreted toxins are a key part of the *Staphylococcus aureus* armoury and many of them produce characteristic symptoms in patients. TSST-1 causes

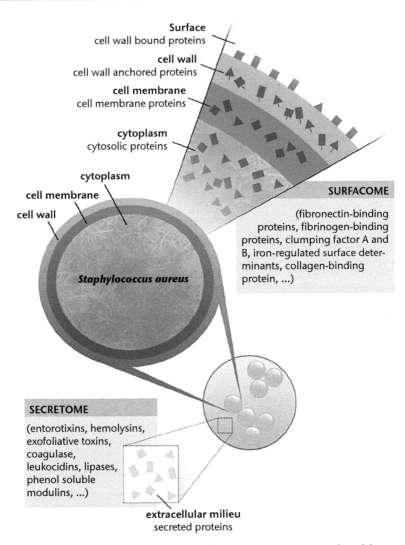

Surface
cell wall bound proteins

cell wall
cell wall anchored proteins

cell membrane
cell membrane proteins

cytoplasm
cytosolic proteins

cytoplasm

cell membrane

cell wall

Staphylococcus aureus

SURFACOME

(fibronectin-binding proteins, fibrinogen-binding proteins, clumping factor A and B, iron-regulated surface determinants, collagen-binding protein, ...)

SECRETOME

(entorotixins, hemolysins, exofoliative toxins, coagulase, leukocidins, lipases, phenol soluble modulins, ...)

extracellular milieu
secreted proteins

Figure 16.2. The virulence factors of *Staphylococcus aureus*. Reproduced by permission of Benedykt Wladyka/*Acta Biochimica Polonica* https://www.researchgate.net/figure/Localization-of-Staphylococcus-aureus-virulence-factors-in-the-context-of-proteomics_fig1_281343548.

toxic shock syndrome which is characterized by fever and multiple organ failure. It was particularly prevalent in the 1970s with toxic shock associated with tampon use. This was due to the introduction of new super-absorbent tampons which favoured the growth of *Staphylococcus aureus*. Some strains of the bacterium produce an enterotoxin that causes gastroenteritis. This form of food poisoning does not result from ingesting the bacteria but rather from ingesting the toxins made by the bacteria that are already present in the contaminated food (Box 16.2). Other strains of *Staphylococcus aureus* produce exfoliative toxins. These are proteases that cause peeling of the skin, a condition seen in children and known as scalded skin syndrome. Panton–Valentine leucocidin is a toxin that destroys white blood cells and causes necrotic lesions involving the skin or mucosa, including necrotic haemorrhagic pneumonia.

As if the above virulence factors were not enough, *Staphylococcus aureus* can act in concert with another skin commensal, the fungus *Candida albicans*. This fungus can cause systemic disease on its own through a morphological switch from a rounded yeast to an invasive hyphal form. Adhesins on the outer surface of *Staphylococcus aureus* can bind to adhesins on the fungal hyphae and enable the bacterium to invade host tissue and avoid host defence mechanisms. In effect, the fungus is acting as a Trojan horse.

Box 16.2 *STAPHYLOCOCCUS AUREUS* FOOD POISONING

Staphylococcal food poisoning is one of the commonest food-borne diseases and results from ingestion of enterotoxins, of which there are at least twenty-two, preformed in food. Symptoms may appear rapidly, sometimes in as little as 30 minutes after consumption of contaminated food but, more usually, it takes up to six hours for symptoms to develop. Outbreaks of this form of food poisoning often are associated with functions such as weddings and other social events resulting in high case numbers presenting to local medical authorities.

Staphylococcus aureus has a high salt tolerance and can grow in ham and other meats, and in dairy products. The toxins that the bacteria produce are heat resistant and cannot be destroyed through cooking. Once food has been contaminated, bacteria begin to multiply. The most common cause of food contamination with *Staphylococcus aureus* is through contact with food workers who are carriers of the bacteria. Foods that require a lot of handling and are stored at room temperature are often the ones involved in staphylococcal food poisoning and include: sandwiches, puddings, cold salads made with mayonnaise, sliced cold meats (particularly if the slicer is not cleaned properly), and cream-filled pastries and desserts.

Comparative genome analysis of a large number of *Staphylococcus aureus* strains has shown that the average genome has 2,800 genes. The core genome consists of 1,441 genes, the accessory genome has 2,871 genes and there are at least 3,145 unique genes. The accessory genome has a large number of genes associated with mobile genetic elements such as transposons and bacteriophages or defence mechanisms. A total of ninety genes for virulence factors have been identified and these are unequally distributed between core and accessory genomes. The number of virulence factors found in each strain ranged from fifty-three to seventy-nine (!) underscoring how *Staphylococcus aureus* is so well developed as a pathogen. Thirty-five genes for virulence factors were found in all strains and these encode the core set of virulence factors whereas the gene for exfoliative toxin B was found in only a single strain.

As noted earlier, β-lactamase-producing *Staphylococcus aureus* arose soon after the introduction of penicillin and by the late 1950s they had become pandemic, especially a clone known as phage-type 80/81. This clone disappeared after the introduction of methicillin but soon was replaced with methicillin-resistant (MRSA) clones. It should be noted that MSRA clones are resistant to almost all β-lactam antibiotics and not just methicillin. Resistance is the result of an additional penicillin-binding protein (PBP, Box 16.1) known as PBP2a that is the product of the *mecA* gene. Counterintuitively, PBP2a differs from other PBPs in *not* binding methicillin or other β-lactam antibiotics. Analysis of penicillin-resistant strains of *Staphylococcus* isolated from patients *before* the introduction of methicillin revealed that some of them were carrying *mecA*, probably as a result of selective pressure from the use of penicillin, but the origin of this gene is not known. Recently, alternative *mec* alleles have been described. For example, *mecC* has 70% sequence identity with *mecA* and is usually found in livestock-associated MRSA (LA-MRSA).

The *mecA* gene is located on a mobile genetic element known as the staphylococcal cassette chromosome (SCC) that ranges in size from twenty-three to sixty-eight kilobase pairs. To date, thirteen variants of SCC have been described. The SCC is important not just for its ability to transfer *mecA* to new hosts but because it also can be the location of resistance determinants to non-β-lactam antibiotics such as fluoroquinolones. This is of clinical importance because the use of fluoroquinolones can drive the prevalence of MRSA.

Initially, MRSA was restricted to hospital-acquired infections but community-acquired MRSA emerged in the 1990s in people who had not visited healthcare facilities. Community-acquired MRSA (CA-MRSA) strains are more virulent than hospital-acquired ones (HA-MRSA) and this may be linked to the Panton–Valentine leucocidin which is a unique feature of the former. Interestingly, although methicillin resistance can be transmitted horizontally, only a limited number of clones are responsible for most of the global cases of MRSA infection. Some clones predominate in geographically limited regions, while others have disseminated worldwide. Using multilocus typing it has been shown that the majority of MRSA infections are caused by strains belonging to a few clonal complexes: CCs 5, 22, and 45 are predominant in hospital settings, while CCs 1, 59, and 80 are frequently isolated in the community. Some, such as CCs 8 and 30 are pandemic in both hospital and community settings.

The way in which horizontal gene transfer contributes to the development of new MRSA strains is shown in Figure 16.3. The *pta*, *yqiL*, and *gmk* genes are three of the seven housekeeping genes used in multilocus typing. In Figure 16.2A, a mutation in the *pta* gene converts an ST30 strain into an ST36 strain. Similarly, in Figure 16.2B, a mutation in *yqiL* converts an ST8 strain into an ST250 strain and an additional *gmk* mutation converts this into an ST247 strain. This figure also shows how methicillin-sensitive strains (MSSA) become MRSA by the acquisition of different versions of the SCC carrying the *mecA* gene. Other virulence genes that were acquired by various strains include those for enterotoxins A, B, K, and Q (*sea*, *seb*, *sek*, and *seq*), Panton–Valentine leucocidin (*pvl*), and TSST (*tst*).

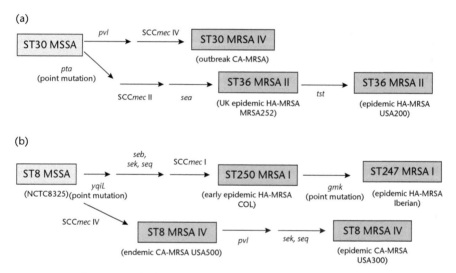

Figure 16.3. The origin of different MRSA strains by horizontal gene transfer. See text for details. Figure reproduced and adapted from Diep B.A., Carleton H.A., Chang R.F., Sensabaugh G.F., and Perdreau-Remington F. (2006) Roles of 34 virulence genes in the evolution of hospital- and community-associated strains of methicillin-resistant *Staphylococcus aureus*. *The Journal of Infectious Diseases* **193**, 1495–503.

Livestock-associated MRSA (LA-MRSA) is a major cause of zoonotic disease in European countries that are major producers of pigs. In Denmark, the incidence of clonal complex CC398 in pigs rose from 3.5% to 90% over a period of ten years. Many farm workers were found to harbour this strain but initially there was little human-to-human spread suggesting that the virulence of the strain had decreased. CC398 strains are descendants of a human variant but in jumping species from humans to pigs there was loss of a prophage known as ΦSa3int. This prophage plays a role in pathogenicity because it carries a group of virulence genes known as the immune evasion cluster. Over time, it was noted that there was increasing human-to-human transmission of the LA-MRSA among the families of farm workers. Then it was noted that this LA-MRSA strain was infecting people who had no contact with farms. Clearly, the virulence of the strain had increased and genomic analysis showed that it had reacquired prophage ΦSa3int by horizontal gene transfer.

As noted earlier, MRSA is a significant cause of healthcare-associated infections and patients need to be treated with antibiotics other than β-lactams. The antibiotic of choice for treatment of MRSA infections in hospital settings is vancomycin, a glycopeptide in clinical use for more than fifty years. In contrast with methicillin, the first report of vancomycin-resistant *Staphylococcus aureus* (VRSA) was not published until thirty-nine years after the drug was first introduced. Low level resistance to vancomycin arises from mutations in genes involved in cell wall biosynthesis whereas high-level resistance is associated with the acquisition of the *vanA* gene. The first report of VRSA carrying the *vanA* gene was in a patient who was co-infected with a vancomycin-resistant strain of *Enterococcus faecalis*. Studies showed that the *vanA* gene was transferred to the *Staphylococcus aureus* on a conjugative plasmid. For reasons that are not understood, VRSA have not become epidemic, and in the twenty years since they were first isolated there have been very few reports of their isolation. Furthermore, all of these VRSA isolates seem to have arisen independently. One possible explanation is that vancomycin resistance weakens the cell wall and makes the bacterium less fit than other bacteria in the absence of selective pressure.

Although vancomycin and some newer antibiotics, such as linezolid and daptomycin, have proved effective in treating MRSA, there is an understandable reluctance by physicians to use these antibiotics. Quite simply, they want to keep them for use as a last resort. The major source of nosocomial MRSA is the patient themselves or the medical staff caring for them. Approximately 5% of people on average are colonized with MRSA and about 20% of them are permanently colonized while the remainder are intermittent carriers. The anterior nares are the most common site of colonization but other body sites are the oropharynx, armpits, perianal region, and groin. Carriage is genetically determined and represents a risk for infection, especially when the carrier is admitted to hospital for surgery. To minimize the risk, most healthcare facilities pre-screen patients for the carriage of MRSA seven to ten days before they are admitted for elective surgery. Patients found to be carrying MRSA are decolonized by nasal administration of the antibiotic mupirocin. This antibiotic has been selected because it only can be given topically and is not used for any other purpose, thereby minimising the development of resistance.

Key points

- *Staphylococcus aureus* is a much-feared pathogen that can cause a number of different diseases depending on the strains involved and the site of entry to the body.
- Staphylococcus has many different virulence factors. Staphyloxanthin (pigment), coagulase, staphylokinase, and the capsule interfere with host defences. Other virulence factors are toxins that cause food poisoning (enterotoxins) or tissue damage (proteases).
- Ninety genes encoding virulence factors have been identified and these are split between the core genome and the accessory genome. Of these genes, thirty-five have been found in all strains.
- Strains that are resistant to methicillin (MRSA) have the *mecA* gene. This is encoded by a mobile genetic element that also harbours other antibiotic-resistance genes. Using ribotyping it has been shown that certain clonal complexes of MRSA predominate worldwide. Community-acquired MRSA are more virulent than hospital-acquired MRSA because they produce the Panton–Valentine leucocidin. Pigs have been shown to be a source of MRSA.
- Vancomycin is one of a small number of antibiotics that can be used to treat MRSA. Resistance to this antibiotic only occurs sporadically. Mupirocin can be used to eliminate nasal carriage of MRSA in patients being admitted to hospital for cold surgery.

Suggested Reading

Chambers H.F. and DeLeo F.R. (2009) Waves of resistance: *Staphylococcus aureus* in the antibiotic era. *Nature Reviews of Microbiology* **7**, 629–41.

Diep B.A., Carleton H.A., Chang R.F., Sensabaugh G.F., and Perdreau-Remington F. (2006) Roles of 34 virulence genes in the evolution of hospital- and community-associated strains of methicillin-resistant *Staphylococcus aureus*. *The Journal of Infectious Diseases* **193**, 1495–503.

Giulieri S.G., Tong S.Y.C., and Williamson D.A. (2020) Using genomics to understand methicillin- and vancomycin-resistant *Staphylococcus aureus* infections. *Microbial Genomics* **6**(1), e000324. doi:10.1099/mgen.0.000324

Sieber R.N., Urth T.R., Petersen A., Møller C.H., Price L.B., et al. (2020) Phage-mediated immune evasion and transmission of livestock-associated methicillin-resistant *Staphylococcus aureus* in humans. *Emerging Infectious Diseases* **26**(1). doi:10.3201/eid2611.201442

Stefani S., Chung D.R., Lindsay J.A., Friedrich A.W., Kearns A.M., et al. (2011) Meticillin-resistant *Staphylococcus aureus* (MRSA): global epidemiology and harmonisation of typing methods. *International Journal of Antimicrobial Agents* **39**, 273–82.

17

The Inadvertent Pathogen: *Borrelia burgdorferi* and Lyme Disease

In 1921, a Swedish dermatologist called Arvid Afzelius published a paper about an unusual rash that he had noted on several patients. This rash was given the name 'erythema chronicum migrans' or ECM because there is a circle of redness (erythema) that moves outwards (migrates) with time. Afzelius speculated that the rash was caused by a bite from an *Ixodes* tick. His work attracted little interest for over fifty years until there was a small outbreak of epidemic arthritis in the United States around the town of Lyme in eastern Connecticut. Of fifty-one patients who were affected, thirteen noted that the disease was preceded by ECM. A subsequent prospective study of thirty-two patients with ECM found that nineteen of them developed arthritic symptoms and three could recall being bitten by a tick at the site of the ECM. Subsequently, ECM was given the common name 'bull's eye' because visually (Figure 17.1) it looks just like the bull's eye on an archery target or dartboard, and the short-lived arthritis was given the name Lyme disease. Attempts were made to isolate causative viruses from ticks but were unsuccessful, which is not surprising given that treatment with penicillin was known to be effective.

Willy Burgdorfer was a Swiss scientist who obtained a PhD degree for his studies on the spirochete (spiral bacterium) *Borrelia duttonii* and its tick vector *Ornithodoros moubata*. In particular, he evaluated the tick's efficiency in transmitting the spirochete during feeding on animal hosts. After completing his PhD, Burgdorfer moved to the United States where he spent many years studying a rickettsial disease, Rocky Mountain Spotted Fever, that is spread by a tick. In October 1981, Burgdorfer received a consignment of *Ixodes dammini* (later renamed *Ixodes scapularis*) ticks from a colleague in Long Island where Lyme disease was endemic. These ticks were not carrying rickettsia but some of them were carrying spirochetes in their midguts. From his PhD studies, Burgdorfer was familiar with the work in Europe on ECM and wondered if

Microbiology of Infectious Disease. Sandy B. Primrose, Oxford University Press.
© Sandy B. Primrose (2022). DOI: 10.1093/oso/9780192863843.003.0017

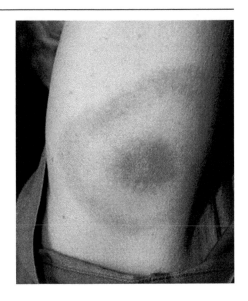

Figure 17.1. The rash known as erythema chronica migrans (ECM). *Source:* James Gathany/Centers for Disease Control and Prevention's Public Health Image Library (PHIL), #9875 (Public domain) <https://commons.wikimedia.org/wiki/File:Erythema_migrans_-_erythematous_rash_in_Lyme_disease_-_PHIL_9875.jpg>.

the spirochetes that he had seen caused Lyme disease. Soon, he had isolated spirochetes from the blood of patients with Lyme disease and patients with ECM, as well as from skin biopsies of cutaneous lesions in patients. Subsequently he showed the presence of spirochetes in the midguts of *Ixodes ricinus*, the European equivalent of *Ixodes dammini*, and in *Ixodes pacificus* which is found in California and Oregon. These spirochetes were given the name *Borrelia burgdorferi* (see Box 17.1).

In nature, *Borrelia burgdorferi* is maintained in an enzootic cycle (Figure 15.2) between an *Ixodes* tick vector and a vertebrate host. The tick larvae acquire the bacteria from infected small mammals such as the white-footed mouse, and occasionally birds, during their first blood meal because transovarial transmission does not occur. The tick will not have another

---

Box 17.1 TERMINOLOGY

The original spirochete isolated by Burgdorfer was found to be similar to the species of *Borrelia* that cause relapsing fever, a disease transmitted by lice and ticks, and was given the name *Borrelia burgdorferi*. Subsequently it was shown that twenty-one species of *Borrelia* can cause Lyme disease but only five of them are significant: *Borrelia burgdorferi*, *Borrelia afzelii* (named after Arvid Afzelius), *Borrelia spielmanii*, *Borrelia bavariensis*, and *Borrelia garinii*. All species that cause Lyme disease are referred to collectively as *Borrelia burgdorferi sensu lato*, while *Borrelia burgdorferi* itself is specified as *Borrelia burgdorferi sensu stricto*.

blood meal until it becomes a nymph in the following Spring. The spirochetes persist in the tick midgut whilst the tick over-winters. Transmission occurs during nymphal feeding in the second year when the first blood meal triggers growth of *Borrelia burgdorferi*. The bacteria migrate to the salivary glands and are injected into the next host on which the tick feeds thereby completing the enzootic cycle. All three stages of ticks feed on humans, which are thought to be incidental hosts, but *Borrelia burgdorferi* transmission by nymphs is considered to cause most cases of Lyme disease.

Different major and minor outer surface proteins of *Borrelia burgdorferi* are made at different stages of the enzootic cycle and represent points of interaction (recognition) between the bacteria and their hosts and vectors. Outer surface protein A (OspA) is synthesized and coats the bacterial surface when spirochetes are acquired by larval ticks feeding on an infected small mammal. This protects the spirochetes from the tick midgut environment, including antibodies against bacterial surface components entering with a blood meal, and it continues to be made as the infected ticks moult to the nymphal stage. When the bacteria migrate from the midgut to the salivary glands, they stop making OspA and start synthesizing another surface protein, OspC, that binds to a tick salivary protein. Successful infection of the mammalian host depends on bacterial expression of OspC. To survive, *Borrelia burgdorferi* must evade the adaptive immune response in its new vertebrate host and it does this using the *vls* (variable major protein-like sequence) system. This is a very sophisticated antigenic variation system that creates highly diverse epitopes of the outer membrane VlsE lipoprotein during *Borrelia*'s time in a mammalian host.

Humans are not part of the enzootic cycle: rather, they are inadvertent hosts and only get bitten if the preferred hosts are absent. Tick bites often go unnoticed because of the small size of the tick as well as tick secretions that prevent the host from feeling any itch or pain from the bite. The number of ticks is directly related to the number of the food sources (mice) for the larvae and nymphs and the number of mice is related to the availability of seeds of deciduous trees. In certain years, known as mast years, the amounts of seed can increase five- to tenfold and the mouse population soars. As humans get bitten in the second year of a tick's life, the incidence of Lyme disease soars in the year after a mast year.

There are some intriguing features of the biology of *Borrelia burgdorferi*. First, the symptoms of the disease vary greatly from mild to very severe (Box 17.2), a variation much greater than seen with other bacterial pathogens. Second, the tick vector in Europe (*Ixodes ricinus*) is different from that in North America (*Ixodes scapularis*). How did this occur? The Atlantic Ocean clearly is a barrier to tick migration and mixing and none of the rodent or deer reservoirs have crossed the Atlantic. Complete genome sequencing

Life Cycle of the *Ixodes scapularis* Tick

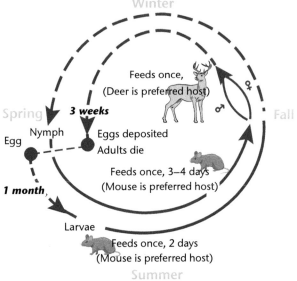

Figure 17.2. Enzootic cycle of *Borrelia burgdorferi*. Spirochetes are acquired when *Ixodes* spp. larvae feed on their first vertebrate host, usually a small mammal or bird. Larvae then moult to nymphs, which transmit the spirochetes when they feed on a second vertebrate host. Nymphs moult to adults, which feed on a third vertebrate host.

Source: Philg88/Wikimedia Commons (Public domain) <https://commons.wikimedia.org/wiki/File:Deer_Tick_life_cycle.svg>.

of many different isolates of *Borrelia burgdorferi* has been done in an effort to try and answer these questions and the results have been surprising.

Box 17.2 SYMPTOMS FOLLOWING *BORRELIA BURGDORFERI* INFECTION

Lyme disease can affect multiple body systems and produce a broad range of symptoms. Not everyone with Lyme disease has all of the symptoms, and many of the symptoms are not specific to Lyme disease but can occur with other diseases as well. Asymptomatic infections are common among those infected in Europe but occur in only 5% of patients in the United States.

Soon after the tick bite, the characteristic bull's eye might appear and it may develop at distant sites as well. Clinical manifestations at this stage can include headache, fever, muscle and joint pain, and general malaise. The second stage, beginning weeks to months after infection, is a result of the hematogenous spread of spirochetes to organs and tissues. It may include arthritis, but the most important features are neurologic disorders which include meningitis and neurologic deficits, and carditis. Other than for arthritis, the pathophysiology of those symptoms is not understood.

The primary component of the cell wall of *Borrelia burgdorferi* is peptidoglycan. The bacteria lack the molecular machinery required for recycling of peptidoglycan during cell replication and so they shed copious amounts of peptidoglycan fragments into the bloodstream. These fragments are recognized by a human pathogen recognition receptor, NOD2, and this stimulates the cells to produce high levels of pro-inflammatory cytokines. This chronic inflammation manifests itself as Lyme arthritis. Synovial fluid from some patients with Lyme arthritis, even those who have received 1–3 months of antibiotic therapy, have high levels of detectible peptidoglycan as well as anti-peptidoglycan antibodies.

Borrelia burgdorferi has the most complex genetic structure identified in any bacterium to date. There is a linear chromosome approximately 900,000 base pairs in length and anywhere between thirteen and twenty-one plasmids. Some of these plasmids are linear and some are circular and they add approximately another 600,000 base pairs to the genome size. Housekeeping genes are constitutively expressed genes that are required for the maintenance of basic cellular functions and in *Borrelia burgdorferi* most of them are on the chromosome. The majority of genes facilitating life in the tick vector and the mammalian host, such as *ospA* and *ospC*, are found on the plasmids. So too are the genes encoding the lipoproteins expressed on the bacterial outer membrane that mediate transition through the enzootic cycle. The importance of lipoproteins to *Borrelia burgdorferi* survival is underscored by their abundance: they represent nearly 8% of all genes. The genes responsible for many of the symptoms of Lyme disease have not been identified but given the differing symptoms between Europe and North America, probably are located on one or more plasmids.

Molecular clock analysis of North American strains of *Borrelia burgdorferi* indicate that the bacterium has been circulating on the continent for at least 60,000 years. Humans only arrived in North America about 24,000 years ago after crossing the Bering Strait and so have no role in the enzootic cycle. The major change came a few centuries ago following colonization by settlers from Europe. This led to deforestation and the widespread killing of predators such as wolves. The deer population increased enormously and with it the ticks that spread Lyme disease. Molecular analysis suggests that trans-Atlantic exchanges of genetic material have happened in the distant past but currently there is no mechanism to account for this. Clearly, we have much to learn about *Borrelia burgdorferi*!

Spirochetes of the genus *Borrelia* are divided into three major groups according to the vector and/or the pathology they can cause. Bacteria of the first group, such as *Borrelia burgdorferi*, cause Lyme disease and are transmitted by hard ticks (*Ixodes* species). Bacteria of the second group, such as *Borrelia duttonii* and *Borrelia hermsii*, are responsible for relapsing fevers and are transmitted by soft ticks (Argasidae). The third group includes species phylogenetically close to species of the second group but which are transmitted by hard ticks. These include *Borrelia theileri* affecting cattle *and Borrelia miyamotoi* affecting rodents. We know very little about these other *Borrelia* species because they have been poorly studied but they must have had a common ancestor and, unsurprisingly, there are many similarities in the symptoms seen in humans. Presumably this ancestor was an obligate parasite that underwent different molecular adaptations as it interacted with different host species and vectors.

Key points

- *Borrelia burgdorferi*, the organism causing Lyme disease, infects humans via tick bites. The European tick vector is different from the North American tick vector. Humans are inadvertent hosts for the ticks which normally feed on wild animals.

- Following infection with *Borrelia burgdorferi*, the symptoms range from none to very severe, the latter including arthritis, neurological disorders, and carditis. Arthritis is caused by release of peptidoglycan cell wall components that stimulate inflammation but the causes of the other symptoms are not known.

- Tick larvae acquire *Borrelia* during their first blood meal. The bacterium over-winters in the tick midgut with the aid of its OspA protein. The first blood meal the following Spring cause the bacterium to stop making OspA and now make the OspC protein. This permits the bacterium to move to the salivary glands and be injected into a new host.

- *Borrelia* avoids the immune system of its mammalian hosts using the vls (variable major protein-like sequence) system to create highly diverse epitopes of the outer membrane VlsE lipoprotein.

- The genome of *Borrelia* is small but very complex as each strain has thirteen to twenty-one plasmids. The majority of genes facilitating life in the tick vector and the mammalian host are found on the plasmids. The genes responsible for many of the symptoms of Lyme disease have not been identified but, given the differing symptoms between Europe and North America, probably are located on one or more plasmids.

Suggested Reading

Barbour A.G. and Benach J.L. (2019) Discovery of the Lyme disease agent. *mBio* **10**(5), e02166–19. doi:10.1128/mBio.02166-19

Brisson D., Drecktrah D., Eggers C.H., and Samuels D.S. (2012) Genetics of *Borrelia burgdorferi. Annual Review of Genetics* **46**, 515–36.

Burgdorfer W. (1984) Discovery of the Lyme disease spirochete and its relation to tick vectors. *The Yale Journal of Biology and Medicine* **57**, 515–20.

Casjens S.R., Lia D. Akther S., Mongodin E.F., Luft B.J., et al. (2018) Primordial origin and diversification of plasmids in Lyme disease agent bacteria. *BMC Genomics* doi:10.1186/s12864-018-4597-x

Castillo-Ramirez S., Fingerle V., Jungnick S., Straubinger R.K., Krebs S., et al. (2016) Trans-Atlantic exchanges have shaped the population structure of the Lyme disease agent *Borrelia burgdorferi* sensu stricto, *Scientific Reports* doi:10.1038/srep22794

Oppler Z.J., O'Keefe K.R., McCoy K.D., and Brisson D. (2020) Evolutionary Genetics of *Borrelia. Current Issues in Molecular Biology* **42**, 97–112.

Walter K.S., Carpi G., Caccone A., and Diuk-Wasser M.A. (2017) Genomic insights into the ancient spread of Lyme disease across North America. *Nature Ecology and Evolution* **1**, 1569–76.

18

Phytoplasmas: Bacteria that Manipulate Plants and Insects

Nearly 200 years ago in Japan it was noticed that the white mulberry trees (*Morus alba*) were afflicted with a disease now known as mulberry dwarf disease. This was of particular concern because the leaves of this tree are the preferred food source of silkworms and silk was a major Japanese export. The disease could be transmitted by grafting of trees and by insects and, following the discovery of viruses, this suggested that the pathogen was a virus. It was 1967 before the true nature of the pathogen was discovered. Electron microscopy revealed the presence of organisms resembling mycoplasmas in the phloem of infected plants (Figure 18.1) but not healthy plants. Phloem is the vascular tissue of plants that carries sugars made in the leaves to other parts of the plant. These mycoplasma-like organisms were given the name phytoplasmas (Box 18.1). Like mycoplasmas, phytoplasmas have no cell wall and so have a variable shape.

Phytoplasmas now are known to be widespread in plants with symptoms described in over 700 species including vegetable and cereal crops, and herbaceous and woody ornamental plants. One characteristic symptom is abnormal flower development. This can take the form of phyllody, the production of leaf-like structures instead of flowers, or virescence where the flowers are green because of loss of pigment (Figure 18.2). Another symptom is leaf yellowing caused by the presence of the bacteria affecting the normal phloem function. Such yellowing has killed millions of coconut palms in the Caribbean. Many infected plants develop a 'witches' broom' appearance (Figure 18.3) because infection triggers a proliferation of axillary (side) shoots and a reduction in the distance between nodes. Witches' broom disease of apples has caused losses running into hundreds of millions of dollars in Europe. Occasionally, however, infection can be turned to advantage: witches' broom is encouraged in commercial production of poinsettias because it gives the plants a much bushier appearance.

Microbiology of Infectious Disease. Sandy B. Primrose, Oxford University Press.
© Sandy B. Primrose (2022). DOI: 10.1093/oso/9780192863843.003.0018

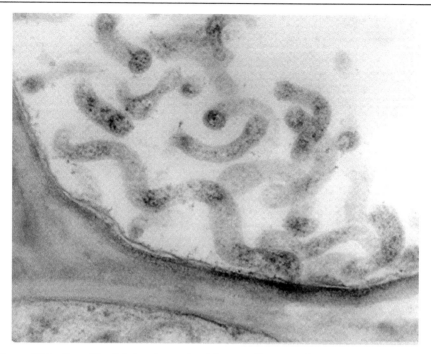

Figure 18.1. Corn Stunt Spiroplasma in phloem cells as seen by transmission electron microscopy.

Source: United States Government/Wikimedia Commons (Public domain) <https://commons. wikimedia.org/wiki/File:Spiro.jpg>.

Figure 18.2. Phyllody of aster flowers. Normal flowers are shown on the right and the flowers of spiroplasma-infected plants are shown on the left.

Source: Amityadav8. Reproduced under Creative Commons Attribution-Share Alike 4.0 International license <https://upload.wikimedia.org/wikipedia/commons/6/61/China_Aster_Phyllody_ Phytoplasma.jpg>.

Figure 18.3. Witches' broom disease of bamboo.

Source: Amityadav8. Reproduced under Creative Commons Attribution-Share Alike 4.0 International license <https://commons.wikimedia.org/wiki/File:Bamboo_Phytoplasma.jpg>.

Transmission of a phytoplasma from a diseased plant to a healthy plant normally occurs via an insect vector such as a leafhopper that feeds on the sap in phloem. Transmission is not mechanical, as happens with many plant viruses, because the phytoplasmas have to multiply within the insect vector. When an insect is feeding on infected plants the phytoplasma penetrates the insect's intestinal wall and circulates in the haemolymph (insect equivalent of blood). The phytoplasmas then migrate to the salivary glands of the insect where they multiply ready to be transmitted to an uninfected plant. Insect transmission is specific between different phytoplasmas and distinct insect vector taxa but phytoplasmas have a wide plant host range and

---

Box 18.1 PHYTOPLASMA TAXONOMY

Unlike the mycoplasmas that infect humans and animals, phytoplasmas cannot be cultured: they are obligate parasites. Because they cannot be cultured, phytoplasmas have been given taxonomic 'Candidatus' status and this leads to cumbersome names such as *Candidatus* Phytoplasma pruni, *Candidatus* Phytoplasma asteris, etc. So far, forty-five Candidatus Phytoplasma species have been named based on analysis of sequence analysis of 16S ribosomal RNA genes. Given that phytoplasmas cannot be cultured, isolating DNA for sequencing needs to be done from infected insects or plant tissue.

can naturally co-infect the same host plants. The insect vectors for the majority of phytoplasmas are unknown or may be difficult to rear in captivity. Fortunately, many phytoplasmas successfully infect Madagascar periwinkle (*Catharanthus roseus*) which can easily be grafted and grown *in vitro*, enabling maintenance of the phytoplasma without the insect vector.

The presence of phytoplasmas may or may not affect the fitness and survival of their insect vectors. Some leafhopper species are negatively impacted by the phytoplasma infection as they die at the time when they are capable of inoculating plants. Other leafhopper species do not show obvious negative effects from infection and sometimes can even benefit from the presence of phytoplasmas by living longer when deprived of a food source and when exposed to suboptimal temperatures. Phytoplasmas also can manipulate plants to become new hosts for leafhoppers that normally do not use these plants as hosts. For example, the leafhopper *Dalbulus maidis* normally will only feed on maize but also will feed on lettuce and China aster plants if these are infected with the Asters Yellow phytoplasma.

An understanding of how phytoplasmas manipulate insects and plants has come from analysis of whole genome sequences of phytoplasmas. The first phytoplasma genome to be sequenced was that of Onion Yellows and it was found to be 860,631 base pairs in length, considerably larger than that of *Mycoplasma genitalium* (580,070 base pairs) which has the minimum genome necessary for a free-living cell. This was surprising as the onion yellows phytoplasma contains even fewer genes than the mycoplasma (Figure 18.4).

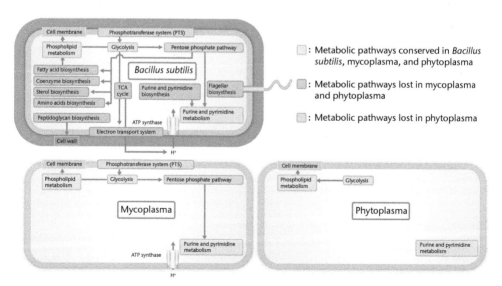

Figure 18.4. Reductive evolution of phytoplasma metabolic pathways. Reproduced with permission from Namba (2019).

The absence of ATP synthase means that phytoplasmas must generate ATP by aerobic respiration. Although metabolic genes are few in number, there are many more transporter genes than in mycoplasmas and these probably reflect the need for phytoplasmas to get nutrients from either insect or plant host cells.

Despite being small, phytoplasma genomes harbour multicopy gene clusters that are named potential mobile units (PMUs). PMU1 is the largest and most complete repeat among the PMUs in the genome of Aster Yellows phytoplasma. It is ~ 20 kilobases in size and contains twenty-one genes encoding DNA replication functions and predicted membrane-targeted proteins. Of particular interest is the observation that there is a linear chromosomal PMU1 and a circular extrachromosomal PMU1. The latter is consistently up to five times higher in copy number in insects compared with plants, and gene expression levels also were considerably higher in insects. This indicates that insects and plants can produce compounds that are transported into phytoplasmas to regulate gene expression.

Phytoplasmas are intracellular parasites that lack a cell wall so their cell membranes play a key role in their interaction with both insects and plants. One group of proteins that comprises the major proportion of total cellular membrane proteins are the phytoplasma immunodominant membrane proteins. They are of three types: immunodominant membrane proteins known as Imps, immunodominant membrane protein A (IdpA), and antigenic membrane protein (Amp). Most phytoplasmas have just one of these protein types although a small number have two of them. The different types do not share any similarity in amino acid sequence but there is homology of sequences within a particular type. The genes for each type of immunodominant protein have specific locations within the genome and they are not allelic. In terms of interaction with insects, the structure and/or composition of the immunodominant membrane protein determines whether the phytoplasma will be transmitted by a particular insect. For example, transmission of Onion Yellows requires that its Amp protein interacts with microfilaments in the guts of its vectors. The immunodominant membrane proteins also interact with plant proteins such as the actin in the plant cytoskeleton.

Proof of the importance of the immunodominant membrane proteins in the infection cycle comes from studies using antibodies. In one study, tobacco plants were engineered to synthesize an antibody to the membrane protein of the stolbur phytoplasma. When grafted on a stolbur phytoplasma-infected tobacco rootstock, the transgenic tobacco shoots grew free of symptoms and flowered after two months. Normal tobacco shoots showed severe stolbur symptoms during the same period and eventually died. In another study, antibodies were used to demonstrate the role of the Amp protein in crossing the insect gut epithelium and salivary gland colonization. These results indicate that immunodominant membrane proteins are suitable targets for developing methods to control phytoplasma infections. Unfortunately, there is a lot of variation within the three types of membrane protein so generating transgenic plants that synthesize the relevant antibodies probably is not the solution.

Whereas phytoplasmas colonize major organs within insects, inside plants they live almost exclusively within the phloem. In some cases, the growth of the phytoplasmas and their sequestering of carbohydrate produced by photosynthesis is sufficient to cause symptoms such as yellowing. However, to induce the developmental and morphological changes that are characteristic of many phytoplasma infections they must secrete effector molecules. These effectors are small peptides and when synthesized by the phytoplasma have the characteristic signal sequence of a protein that will be exported. Also, the small size of these effectors means that they can be distributed systemically and reach tissues distant to the phloem.

The first effector to be characterized is one known as TENGU. This is produced by many phytoplasma strains and is a secreted peptide of thirty-eight amino acids. The TENGU peptide is cleaved by the plant to an oligopeptide of just twelve amino acids. This gets transported to the apical meristem where it interferes with auxin production and thus down-regulates many auxin-related genes. Auxin normally is produced at the shoot apex and is transported down the stem to inhibit shoot branching but in its absence the plant develops witches' broom. Another effector, SWP1 also induces witches' broom but by a different mechanism. It interacts with a key plant regulator, BRC1 that prevents branching, and facilitates its degradation.

Another class of effectors, such as SAP54, are known as phyllogens because they induce virescence, phyllody, and proliferation symptoms in flowers. SAP54 exerts its effect by promoting the degradation of proteins that regulate important developmental processes in flowering plants. These proteins are highly conserved transcription factors and reducing their activity through SAP54-mediated degradation curtails flower development, generating sterile plants. Many different phyllogens have been identified in phytoplasmas but their origin is unclear because they are not homologous with any other protein sequences. Phyllogen genes are associated with potential mobile units (PMUs) suggesting that horizontal gene transfer has contributed to the acquisition and sharing of phyllody-inducing activity among phytoplasmas.

Why do phytoplasmas induce symptoms accompanied by unique morphological changes such as witches' broom and phyllody? Both symptoms increase the prevalence of short branches and small young leaves, which are preferred by sap-feeding insects. Furthermore, the life of small young leaves and flowers with phyllody symptoms is prolonged. In particular, phyllody flowers remain green even when healthy flowers wither. These features are likely to enhance attraction of insect vectors and thus the spread of phytoplasmas. Also, effectors such as TENGU and SWP1 down-regulate the production of the plant hormone jasmonic acid in addition to auxin. In the absence of jasmonic acid, which is a volatile insect repellent, the plants are a lot more attractive to insect vectors. Such manipulations of the morphology and biochemistry of host plants appear to be a common strategy for the survival of phytoplasmas.

Key points

- Phytoplasmas are mycoplasmas that infect plants and insects. In plants they cause symptoms such as abnormal flower development (virescence, phyllody) and 'witches' broom'.

- Insects acquire phytoplasmas when they feed on infected plants. After ingestion, the phytoplasmas infect the haemolymph and then migrate to the salivary glands ready to be injected into another plant. Different phytoplasmas infect different vectors and different vectors feed on different plants. Infection with a phytoplasma can affect the behaviour of the insect.

- The phytoplasma genome is about 50% bigger than that of *Mycoplasma genitalium* but has fewer genes and so phytoplasmas are dependent on their host plants and insects for essential nutrients. Phytoplasmas contain multicopy gene clusters called Potential Mobile Units (PMUs), some of which are on plasmids. The copy number of these PMUs differs in plants and animals showing a host effect on the bacteria.

- Phytoplasmas have no cell wall and so membrane proteins modulate the interaction with hosts. There are three types of immunomodulatory protein (Imp) found on the membrane. They are non-allelic but most phytoplasmas have just one. The Imp that a phytoplasma has determines which insects can be its vector.

- In plants, phytoplasmas are restricted to the phloem and so the disease symptoms are produced by small molecule effectors. Some of these effectors are encoded by PMUs, suggesting acquisition by horizontal gene transfer, but their origin is not known.

Suggested Reading

Iwabuchi N., Kitazawa Y., Maejima K., Koinuma H., Miyazaki A., et al. (2020) Functional variation in phyllogen, a phyllody-inducing phytoplasma effectorfamily, attributable to a single amino acid polymorphism. *Molecular Plant Pathology* **21**, 1322–36.

Konnerth A., Krczal G., and Boonrod K. (2016) Immunodominant membrane proteins of phytoplasmas. *Microbiology* **162**, 1267–73.

Namba S. (2019) Molecular and biological properties of phytoplasmas. *Proceedings of the Japanese Academy, Series B* **95**, 401–18.

Wang N., Yang H., Yin Z., Liu W., Sun L., and Wu Y. (2018) Phytoplasma effector SWP1 induces witches' broom symptom by destabilizing the TCP transcription factor BRANCHED. *Molecular Plant Pathology* **19**, 2623–34.

Wei W., Trivellone V., Dietrich C.H., Zhao Y., Bottner-Parker K.D. & Ivanuskas A. (2021) Identification of phytoplasmas representing multiple new genetic lineages from phloem-feeding leafhoppers highlights the diversity of phytoplasmas and their potential vectors. *Pathogens* **10**(3), 352. doi: 10.3390/pathogens10030352

19

The Most Influential Bacterium: *Wolbachia pipientis*

Typhus is a disease caused by bacteria belonging to the genus *Rickettsia* that are transmitted by fleas, lice, and ticks. Knowing this, in 1924 Marshall Hertig and Simeon Wolbach began screening arthropods found in the area around Boston for the presence of *Rickettsia*-like bacteria. In the mosquito *Culex pipiens* they found bacteria residing within male and female reproductive cells (Figure 19.1) but absent in all other cells. When mosquitos carrying the bacteria were reared in the laboratory, their offspring also harboured them in their reproductive tissues suggesting that the bacteria were maternally transmitted. Hertig later named the bacteria *Wolbachia* in honour of Wolbach who had been his PhD supervisor.

In 1971, Janice Yen and Ralph Barr announced that they had found the answer to a problem that had long puzzled biologists: what was the cause of embryonic death when *Culex* mosquitos from different geographic areas were mated? The phenomenon is known as cytoplasmic incompatibility (CI) and it is caused by the presence of *Wolbachia*. CI occurs when a *Wolbachia*-infected male mates with a female that is infected with another strain of *Wolbachia* (bidirectional CI) or is uninfected (unidirectional CI). An infected female is only compatible with a male infected with the same strain of *Wolbachia* and an uninfected female is only compatible with an uninfected male (Table 19.1). These observations have a significant evolutionary implication. As *Wolbachia* are transmitted only by females, CI promotes the spread of *Wolbachia* and keeps it from dying out. This explains the high incidence (40–60%) of *Wolbachia* in insects and other arthropods and it is thought that there are more than a million infected host species worldwide! As we shall see later, this observation also has implications for biological control.

Despite the role of *Wolbachia* in CI being known for over fifty years, the mechanism whereby it occurs still remains a mystery even though many *Wolbachia* genomes have been sequenced. However, the sequencing of a *Wolbachia* genome (wMel) from *Drosophila melanogaster*, coupled with the

Microbiology of Infectious Disease. Sandy B. Primrose, Oxford University Press.
© Sandy B. Primrose (2022). DOI: 10.1093/oso/9780192863843.003.0019

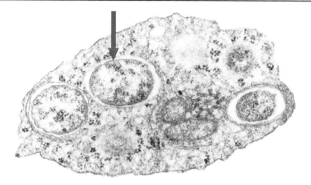

Figure 19.1. *Wolbachia* (shown by arrow) growing inside a reproductive cell.

Source: Scott O'Neill/Genome Sequence of the Intracellular Bacterium Wolbachia. *PLoS Biol* 2/3/2004: e76. doi:10.1371/journal.pbio.0020076. Reproduced under Creative Commons Attribution 2.5 Generic license <https://commons.wikimedia.org/wiki/File:Wolbachia.png>.

Table 19.1. Viable and unviable crosses between *Wolbachia*-infected mosquitoes.

Wolbachia strain in male insect	Wolbachia strain in female insect	Outcome for progeny
A	B	Death
A	none	Death
A	A	Survival
B	B	Survival
none	none	Survival

extensively studied genetics of the fruit fly has allowed some progress. The genome is about 1.2 Mb in size and, being from an endosymbiont, has lost many of the genes involved in cell wall biosynthesis. Surprisingly for an intracellular bacterium, and one with such a small genome, it has very large amounts of repetitive DNA and DNA corresponding to mobile genetic elements. Notable among these elements is prophage WO that encodes a set of proteins termed the Eukaryotic Association Module. These proteins share homology with proteins from eukaryotes and *Wolbachia* probably acquired them by horizontal gene transfer. The wMel genome also is enriched for ankyrins that are involved in protein–protein interactions in eukaryotes but not free-living bacteria. The link between phage WO and/or ankyrin genes and CI was made when both were found to be absent from a strain of *Wolbachia* that did not manipulate host reproductive fitness.

Analysis of the genomes from several different *Wolbachia* genomes led to the identification of two genes, *cifA* and *cifB*, in the Eukaryotic Association Module that play a role in CI. Structural homology-based analyses suggest that the CifA protein has three domains. Although the putative function of these domains can be identified, how the protein functions in *Wolbachia*-infected organisms is not known. The role of the CifB protein also is anomalous. The CifA and CifB proteins can bind to each other *in vitro* and to a suite of host proteins but the *in vivo* interactions remain to be elucidated. Given that *Wolbachia* has the ability to manipulate its host, it must have a mechanism for secreting effectors such as the Cif proteins into host cells. Therefore, it should come as no surprise that the *Wolbachia* genome encodes a type IV secretion system.

It now is known that *Wolbachia* can manipulate reproduction in insects in ways other than CI. In some flies, *Wolbachia* increases fecundity such that infected flies produce more offspring than uninfected ones. In other insects, male killing occurs when infected males die during larval development resulting in a greater number of infected females. In other insects, infected males become feminized. Finally, infected females sometimes can reproduce in the absence of males (parthenogenesis). These observations are particularly important since many diseases are spread by the females of biting insects, such as mosquitos, because they must ingest vertebrate blood before laying eggs otherwise the eggs will not mature.

Not all infections with *Wolbachia* result in reproductive manipulation of the host. Strains that have no effect on reproductive fitness are still widespread indicating that they have other beneficial effects for their hosts. In some arthropods, *Wolbachia* provides essential nutrients whereas in others it provides an increased resistance to insecticides. In certain leafminers, the presence of *Wolbachia* help their hosts to produce green islands on yellowing tree leaves so that the larvae can grow to an adult form. This effect is eliminated if the larvae are treated with antibiotics. When infected with *Wolbachia*, the fruit fly *Drosophila melanogaster* is more resistant to the RNA viruses Drosophila C virus, Nora virus, and Flock House virus, but not to a DNA virus (Insect Iridescent Virus). By reducing the load of viruses, *Wolbachia* increases the fitness of infected flies.

Wolbachia strains are not restricted to insects. They infect a wide variety of arthropods including isopods (e.g. woodlice), mites, and spiders but they also can infect crustaceans and nematodes. Of particular significance is the presence of *Wolbachia* in the filarial worms *Brugia malayi* and *Wuchereria bancrofti*. These worms are responsible for the disease known as *elephantiasis tropica*, which is a gross enlargement of infected body parts (Figure 19.2), and which affects about 120 million people in Africa and south-east Asia. *Wolbachia* is an obligate symbiont and provides the host worm with essential nutrients. The worms are transmitted by mosquitos, but in a twist of fate, if the mosquitos are carrying *Wolbachia* their worm-carrying ability is reduced. Prior to the discovery of the role of *Wolbachia* in elephantiasis, patients were treated with anti-nematode medications which are very toxic. Now they can be treated simply with the antibiotic doxycycline and the World Health Organization is using this in a global elimination programme.

Figure 19.2. Elephantiasis of the leg of a Filipino man due to filariasis.
Source: Centers for Disease Control and Prevention's Public Health Image Library (PHIL), #373 (Public domain). <https://commons.wikimedia.org/wiki/File:Elephantiasis.jpg>.

Another human disease where *Wolbachia* plays a part is onchocerciasis, also known as river blindness. This is a disease caused by infection with the parasitic worm *Onchocerca volvulus* and it is spread by the bites of a black fly. This fly lives near rivers, hence the common name of the disease. *Wolbachia* is an endosymbiont of adult worms and microfilariae and is thought to be the driving force for morbidity. Dying microfilariae release *Wolbachia* surface proteins that trigger innate immune responses and produce inflammation and its associated morbidity. People with river blindness are treated with ivermectin, which kills the larvae but not the adult worms, and doxycycline which weakens the worms by killing *Wolbachia*.

Mosquitos are responsible for the transmission of a large number of viruses that cause human disease including dengue, chikungunya, yellow fever, Zika, and West Nile fever. The global health and economic burden from these diseases is incalculable. Conventional efforts to control these diseases are based either on elimination of the insect vector or vaccination of susceptible individuals but cost considerations mean that both approaches will have limited success. *Wolbachia* offers not one, but two, viable alternatives. The first of these is *population suppression* (Figure 19.3) to reduce the population size of disease vectors by rearing CI-inducing male insects in the laboratory and releasing them into the environment. This is a variation of an established technique in which male insects are sterilized with X-rays before release. The first large-scale test of using CI-inducing males was undertaken in Fresno, California in 2018 and achieved a 95% reduction in female mosquitos.

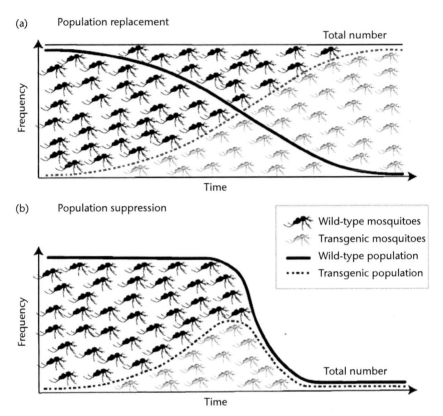

Figure 19.3. Comparison of population replacement (A) and population suppression (B) strategies. Reproduced by permission of Hector Quemada <https://www.researchgate.net/figure/Comparison-of-population-replacement-A-and-population-suppression-B-strategies_fig1_325645812>.

The second method of using *Wolbachia* in disease control is known as the *population replacement strategy*. In this case, the aim is not to reduce the size of the mosquito vector population but to convert it to one that has a reduced capacity to transmit viruses. As noted earlier, *Wolbachia* can increase the fitness of fruit flies by increasing their resistance to various viruses. In a similar fashion, *Wolbachia* also is able to inhibit the replication of a wide variety of RNA viruses in mosquitos although the mechanism whereby this occurs is not known. This population replacement strategy has been used twice in large-scale trials, once in Indonesia and once in Australia. In Townsville, Australia, there were no reports of dengue for the first time in over thirteen years after *Wolbachia* was introduced to the local population of mosquitos, and in the Indonesia city of Yogyakarta there was a 77% reduction in cases of dengue. It is worth noting that the population replacement strategy is not a panacea: in some cases, *Wolbachia* infection leads to increased transmission of viruses.

Key points

- *Wolbachia* is an endosymbiont of arthropods that can manipulate their reproductive fitness via cytoplasmic incompatibility. This phenomenon has been used successfully to reduce the transmission by mosquitoes of viral diseases.

- Genome analysis has shown that *Wolbachia* has shed many genes for cell wall biosynthesis but has large amounts of repetitive DNA and DNA of mobile genetic elements. Prophage WO encodes the Eukaryotic Association Module that is the site of many genes for eukaryotic-like proteins. Other parts of the genome also encode eukaryotic-like proteins but the role of these proteins is not understood. The genome encodes a T4SS that permits transfer of effectors to the insect host.

- *Wolbachia* can manipulate its hosts in ways other than reproductive fitness, for example, increased fecundity, feminization, and resistance to insect viruses.

- *Wolbachia* can infect invertebrates other than arthropods. *Wolbachia*-infected nematodes cause elephantiasis and infected filarial worms cause river blindness. Both diseases can be successfully treated by using antibiotics to kill *Wolbachia*.

Suggested Reading

Crawford J.E., Clarke D.W., and White B.J. (2020) Efficient production of male *Wolbachia*-infected *Aedes aegypti* mosquitoes enables large-scale suppression of wild populations. *Nature Biotechnology* **38**, 482–92.

Newton I.L.G. and Rice D.W. (2020) The Jekyll and Hyde symbiont: could *Wolbachia* be a nutritional mutualist? *Journal of Bacteriology* **202**, e00589-19. doi:10.1128/JB.00589-19

Newton I.L.G. and Slatko B.E. (2019) Symbiosis comes of age at the 10th biennial meeting of *Wolbachia* researchers. *Applied and Environmental Microbiology* **85**(8), e3071–18. doi: 10.1128/AEM.03071-18

O'Neill S.L., Ryan P.A., Turley A.P., Wilson G., Retzki K. et al. (2018) Scaled deployment of *Wolbachia* to protect the community from dengue and other *Aedes* transmitted arboviruses. *Gates Open Research* **2**, 26. doi:10.12688/gatesopenres.12844.3

Pimentel A.C., Cesar C.S., Martins M., and Cogni R. (2021) The antiviral effects of the symbiont bacteria *Wolbachia* in insects. *Frontiers in Immunology* **11**, 626329. doi:10.3389/fimmu.2020.626329

Shropshire J.D., Leigh B., and Bordenstein S.R. (2020) Symbiont-mediated cytoplasmic incompatibility: what have we learned in 50 years. *eLife* **9**, e61989. doi:10.7554/elife.61989

Part III
Eukaryotic Pathogens

20

The Ubiquitous Pathogen: *Trichomonas vaginalis*

According to estimates by the World Health Organization, in 2016 there were an estimated 110 million individuals worldwide with infections due to the protozoan *Trichomonas vaginalis*—three times more than those living with HIV/ AIDS, although the true incidence could be much higher. According to the Centers for Disease Control, 2.1% of US women aged fourteen to fifty-nine are infected, with the incidence being as high as 9% in African-American women. Despite these appalling statistics, the disease is little known outside clinics specializing in sexually transmitted diseases. Consequently, *Trichomonas vaginalis* has been much less studied than much rarer pathogens even though infection with it doubles the chances of being infected with HIV. Why is this? Possibly it is because the symptoms are not particularly overt: vaginal pruritis and discharge in women and few symptoms, if any, in men. However, the symptoms can be more serious. Women can develop inflammation of the cervix, urethritis, and adverse pregnancy outcomes (premature delivery, low birth weight, mortality) and men can get urethritis and prostatitis.

Being a protozoan, *Trichomonas vaginalis* is classified as a eukaryotic organism and, as such, would be expected to have mitochondria. However, living as it does as an obligate pathogen in the vagina and urethra, it has little exposure to oxygen and so has no need for normal mitochondria. This means that it lacks the enzymes and cytochromes necessary to conduct oxidative phosphorylation. Instead, of mitochondria, *Trichomonas* has hydrogenosomes which have evolved from mitochondria by extensive gene loss. Like mitochondria, hydrogenosomes are bounded by a double membrane but they lack a genome and so all their proteins are encoded by the nuclear genome. Hydrogenosomes contain the enzymes pyruvate-ferredoxin oxido-reductase and hydrogenase which allows them to generate ATP from pyruvate and malate whilst producing acetate and hydrogen gas (Figure 20.1). The hydrogenosome is of importance clinically because the

Microbiology of Infectious Disease. Sandy B. Primrose, Oxford University Press.
© Sandy B. Primrose (2022). DOI: 10.1093/oso/9780192863843.003.0020

pyruvate-ferredoxin oxido-reductase that it contains activates the antibiotics metronidazole (Box 20.1) and tinidazole. Because this reduction happens only in anaerobic bacteria and protozoans such as *Trichomonas*, the antibiotics have relatively little effect upon human cells or aerobic bacteria forming the normal microflora.

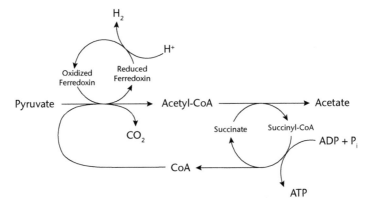

Figure 20.1. The production of ATP from pyruvate with the concomitant production of acetate and hydrogen gas.

The most common microbial vaginal imbalance is bacterial vaginosis and the most common cause is infection with *Gardnerella vaginalis*. However, other bacteria can cause bacterial vaginosis and one of these is the obligate pathogen *Mycoplasma hominis*. Interestingly, although both *Mycoplasma hominis* and *Trichomonas vaginalis* can cause vaginal disease independently, they can act symbiotically with the mycoplasma living internally in the trichomonad. Furthermore, the mycoplasmal infection can be passed from mycoplasma-infected trichomonads to mycoplasma-free trichomonads and to human cervical cells. The presence of *Mycoplasma hominis* within *Trichomonas vaginalis* has been shown to protect the former from antibiotics and the host immune system whilst enhancing the growth rate and pathogenicity of the latter.

In addition to *Mycoplasma hominis*, many strains of *Trichomonas vaginalis* are persistently infected with small (5 Kb) double-stranded RNA viruses called *Trichomonas vaginalis* viruses (TVVs). TVVs affect the total protein expression of the host and especially the expression of proteins which are linked to cytotoxicity, cytoadherence, and host immune evasion. Because TVVs contain double-stranded RNA they can trigger the production of type I interferon by the human host as well as a number of proinflammatory mediators and chemokines implicated in the pathogenesis of trichomoniasis. TVVs can be divided into four different viral strains that have the ability to co-infect *Trichomonas vaginalis* at the same time: TVV1, TVV2, TVV3, and TVV4. Each TVV strain has different effects on various aspects of tri-chomonad pathogenesis. TVV1 and TVV2 have been linked to genital symp-tom severity, while TVV2 and TVV3 are involved in the surface expression of virulence factors of *Trichomonas vaginalis*. The role of TVV4 has yet to be elucidated.

The chance of *Trichomonas vaginalis* successfully colonizing a new host depends on the bacterial flora with which it has to compete. Lactobacilli are the key to a healthy vagina. Not only do they generate a pH value of 2.8–4.2 through the production of lactic acid, they strongly inhibit adhesion of *Trichomonas vaginalis* to epithelial cells. The lactobacilli even can displace trichomonads that have attached to cells. However, the trichomonad can fight back. It is known to produce nine peptidoglycan hydrolases in response to the presence of bacteria. In laboratory experiments, these enzymes have been shown to enable *Trichomonas vaginalis* to kill bacteria with which they are co-cultured by weakening their cell wall. If a similar killing effect is seen against lactobacilli in the vagina, then it would explain why the vaginal pH value rises above 4.5 when there is a *Trichomonas vaginalis* infection. Since mycoplasmas do not have a cell wall, *Mycoplasma hominis* is unaffected by the peptidoglycan hydrolases.

Trichomonas vaginalis has flagella and so can swim through the vaginal mucus until it comes in contact with epithelial cells. Contact induces a cytoskeletal rearrangement resulting in the rapid morphological transition from spindle-shaped flagellates to tissue-feeding and actively dividing amoe-boids. This change to an amoeboid shape results in a greater surface area for cell-to-cell contact. *Trichomonas vaginalis* is an extracellular parasite that does not become internalized by the host cell. Instead, it adheres to host urogenital epithelial cells and causes their lysis. Infection is facilitated by the production and shedding by *Trichomonas vaginalis* of extracellular vesicles. These vesicles are of two types: exosomes derived from intracellular multi-vesicular bodies and microvesicles derived from the cell membrane. They contain various proteins and RNA whose role in infection is not fully un-derstood. What is known is that these extracellular vesicles fuse with the epithelial cells and deliver their contents into them thereby modulating a host cell immune response.

The genome of *Trichomonas vaginalis* has been sequenced and at 160 Mb it is the largest protist genome sequenced to date. The complete genome sequence has not been assembled because ~ 60% of the genome consists of repeated sequences. Despite this, a lot of useful information has been gleaned. A surprisingly large portion of the genome consists of transpos-able elements: there are about 40,000 transposable element genes distributed among around fifty-nine transposable element families and members of each family show a high degree of homology. The genome has about 60,000 genes encoding proteins, many of them members of high copy number gene families.

Among the expanded gene families are some that encode surface proteins thought to mediate adhesion to epithelial cells. BspA (Bacteroides surface protein A) and Pmp (polymorphic membrane protein) are surface adhesion molecules known to be involved in other host parasite interactions. *Trichomonas vaginalis* encodes 911 BspA-like and 48 Pmp-like proteins. Analysis of the genes encoding these proteins suggest that they may have been acquired by horizontal gene transfer from *Bacteroides*-related bacteria. These bacteria are abundant in the flora of the large intestine, which is also the home of most trichomonads, and transfer could have occurred there before *Trichomonas vaginalis* transitioned to the urogenital tract.

Adherence of *Trichomonas vaginalis* is followed by lysis of human cells including vaginal or prostate epithelial cells and red or white blood cells. The parasite genome encodes many enzymes that could do this, including over 400 proteases, but identifying the ones most relevant to pathogenicity has proved difficult. One group of proteins that could be of relevance are saponin-like pore-forming proteins (TvSaplips). The synthesis of one of these, TvSaplip12, is increased significantly on contact with host cells and it has a strong lytic activity against them. It also can lyse bacteria so possibly plays a role in allowing *Trichomonas vaginalis* to eliminate competition from other members of the vaginal flora. When the activities of the parasite result in mucosal damage, host toll-like receptors recognize specific features of *Trichomonas vaginalis* (the pathogen-associated molecular patterns or PAMPs). This recognition leads to activation of various inflammatory pathways and the characteristic symptoms experienced by the patient. One factor in the broad spectrum of disease severity associated with trichomoniasis is likely to be the strain of parasite, because clinical isolates vary broadly in their ability to kill cervicovaginal and prostate epithelial cells *in vitro*. However, host factors such as an individual's microbiome and strength of immune response most likely play a role as well.

For over sixty years, metronidazole has been the antibiotic of choice for treating *Trichomonas vaginalis* infections but the incidence of resistance in clinical isolates is approaching 10%. Some of the mechanisms of resistance are common to many antibiotics. For example, some resistant strains have increased amounts of efflux pumps that will export the drug from the cell. Another resistance mechanism is inactivation of metronidazole by a reductase known as Nim2. In *Bacteroides* species, Nim genes are carried on insertion sequences that can be acquired by horizontal gene transfer and this might be the case in *Trichomonas vaginalis*. Other resistance mechanisms are related to the activation of the drug by the hydrogenosome. For example, mutations in oxygen stress response genes such as those encoding NADH oxidase and flavin reductase, lead to elevated levels of intracellular oxygen and reoxidation of the nitro free radical form of metronidazole. Other mutations that have been observed are ones in genes for the pyruvate-ferredoxin oxido-reductase.

Box 20.1 MODE OF ACTION OF METRONIDAZOLE

Metronidazole (Figure 20.2) is the drug of choice for treating infections caused by obligate anaerobic organisms such as *Clostridium difficile* (chapter 15) and *Trichomonas vaginalis*. It is a prodrug and it is activated by reduction of its nitro group. Pyruvate-ferredoxin oxido-reductase normally generates adenosine triphosphate (ATP) via oxidative decarboxylation of pyruvate. With metronidazole in the cellular environment, its nitro group acts as an electron sink, capturing electrons that would usually be transferred to hydrogen ions in this cycle. The first step of the reductive activation of metronidazole is proposed to form the nitro free radical, followed by the nitroso free radical and hydroxylamine derivatives.

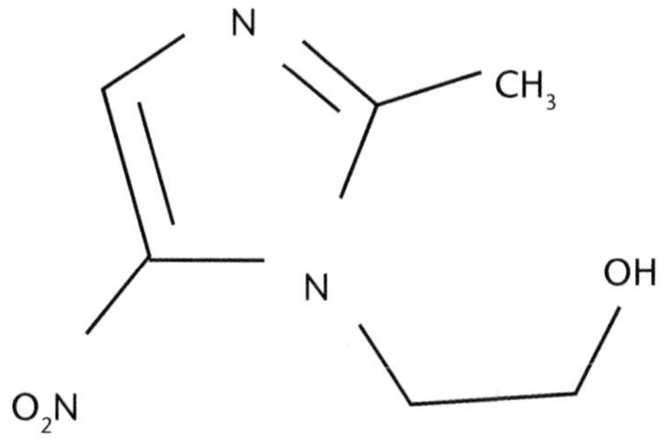

Figure 20.2. The structure of metronidazole.

Reduction of metronidazole creates a concentration gradient that drives uptake of more drug and promotes formation of intermediate compounds and free radicals that are toxic to the cell. Determining how the metabolites of metronidazole exert cytotoxicity after reductive activation is challenging, as intermediates are short-lived. However, it generally is believed that they interact with DNA and inhibit replication and repair.

Sexual reproduction is a major source of genetic diversity in populations of higher plants and animals. The advantage of it is that it accelerates adaptation to fluctuating environments and helps to remove deleterious mutations. In asexual populations, all cells carry the same genetic information and new combinations of genes arise from mutation and horizontal gene transfer. *Trichomonas vaginalis* clearly is a very successful parasite but does this success have its origins in sexual reproduction? There are two criteria for determining if sexual reproduction plays a key role in the life of *Trichomonas vaginalis*: are there substantially different populations in the many clinical isolates of the pathogen, and can it undergo meiosis? To date, only two different global subpopulations have been identified which is much lower than would be expected if there was sexual reproduction. On the other hand, genomic analysis has shown that the pathogen has twenty-seven of the twenty-nine necessary meiosis genes and all are functional. So, like many aspects of the biology of *Trichomonas vaginalis*, the question about sexual reproduction remains unanswered.

Key points

- *Trichomonas vaginalis* is an obligate pathogen that infects over 100 million people annually. As it lives in anaerobic environments it does not have mitochondria and eliminates excess electrons by forming hydrogen in an organelle known as the hydrogenosome.

- *Mycoplasma hominis* can exist symbiotically inside *Trichomonas vaginalis* to the mutual benefit of both partners.

- In order to colonize the vagina, *Trichomonas* needs to raise the pH and it does this by killing off lactobacilli and other acid-producing bacteria by secreting peptidoglycan hydrolases.

- *Trichomonas vaginalis* adheres to the epithelium of the vagina and urinary tract but is not internalized. Instead, it sheds exosomes and microvesicles which fuse with epithelial cells and release a range of protein and RNA effectors. The net effect is epithelial cell lysis and inflammation which manifest themselves in the disease symptoms.

- Genome sequencing has revealed that *Trichomonas* has a very large genome (160 Mb) containing many repeated sequences and fifty-nine families of transposable elements. There a very large number (~ 1000) of genes for surface adhesion molecules and many different virulence factors have been identified including 4 RNA viruses.

- As *Trichomonas vaginalis* is an anaerobe, the antibiotic of choice is metronidazole, especially as this will have little effect on the normal flora. However, resistance to metronidazole is increasing but no transmissible elements are involved.

Suggested Reading

Bradic M. ands Carlton J.A. (2018) Does the common sexually transmitted parasite *Trichomonas vaginalis* have sex? *PLoS Pathogens* **14**, e1006831. doi:10.1371/journal.ppat.1006831

Bradic M., Warring D., Tooley G.E., Scheid P., Secor W.E. et al. (2017) Genomic indicators of drug resistance in the highly repetitive genome of *Trichomonas vaginalis*. *Genome Biology and Evolution* **9**, 1658–72.

Leitsch D. (2021) Recent advances in the molecular biology of the protist parasite *Trichomonas vaginalis*. *Faculty Opinions* **10**, 26. doi:10.12703/r/10-26

Lewis W.H., Lind A.E., Sendra K.M., Onsbring H., Williams T.A., et al. (2020) Convergent evolution of hydrogenosomes from mitochondria by gene transfer and loss. *Molecular Biology and Evolution* **37**, 524–39.

Margarita V., Fiori P.L., and Rappelli P. (2020) Impact of symbiosis between *Trichomonas vaginalis* and *Mycoplasma hominis* on vaginal dysbiosis: a mini review. *Frontiers in Cellular and Infection Microbiology* **10**, 179. doi:10.3389/fcimb.2020.00179

Margarita V., Rappelli P., Dessi D., Pintus G., Hirt R.P., and Fion P.L. (2016) Symbiotic association with *Mycoplasma hominis* can influence growth rate, ATP production, cytolysis and inflammatory response of *Trichomonas vaginalis*. *Frontiers in Microbiology* **7**, 953. doi:10.3389/fmicb.2016.0095

21

The Greatest Killer of All Time: The Malarial Parasite

In human history, malaria has caused more deaths by far than any other infectious disease. Today, with our understanding of the biology of the parasite, we know that hundreds of millions of people are infected annually and many of them will die from the disease. Malaria is due to a blood infection by protozoan parasites of the genus *Plasmodium* which are transmitted from one human to another by female mosquitos, most commonly of the genus *Anopheles*. Five species of *Plasmodium* can infect humans: *Plasmodium falciparum*, *Plasmodium vivax*, *Plasmodium ovale*, *Plasmodium malariae*, and *Plasmodium knowlesi*. Most deaths, primarily in young children, are caused by *Plasmodium falciparum* which is highly prevalent in sub-Saharan Africa. *Plasmodium vivax* is rare in sub-Saharan Africa but endemic in many parts of Asia, Oceania, and Central and South America. It also is found sporadically in some temperate regions where once it was widely prevalent, for example in Europe. Between them, *Plasmodium falciparum* and *Plasmodium vivax* account for 95% of all human malarial infections.

The malarial parasite has a complicated life cycle involving replication in both the human host and the mosquito vector (Figure 21.1). The infection cycle begins when a person is bitten by an infected female mosquito in search of a blood meal. Malarial sporozoites are inoculated into the dermis and enter the bloodstream where they are carried to the liver. Here the parasites enter hepatic cells and grow, divide, and develop into merozoites. Merozoites are released into the bloodstream where they invade and multiply in red blood cells. After several erythrocytic cycles, and possibly stimulation from the immune system, male and female gametocytes form, and these gametocytes are infective for mosquitos. When an uninfected mosquito takes a blood meal from an infected patient, it ingests the gametocytes and these develop into gametes which mate to produce zygotes. The zygotes develop into ookinetes that penetrate the mosquito midgut wall and develop into oocysts. These in

Microbiology of Infectious Disease. Sandy B. Primrose, Oxford University Press.
© Sandy B. Primrose (2022). DOI: 10.1093/oso/9780192863843.003.0021

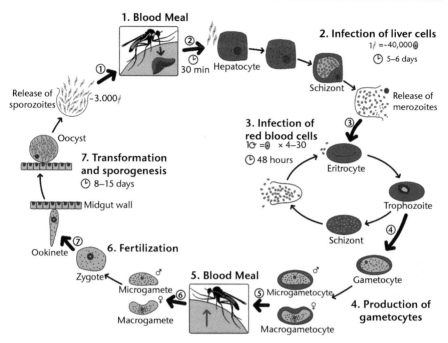

Figure 21.1. The *Plasmodium* life cycle. A malaria infection begins with the transmission of a *Plasmodium* parasite via a female *Anopheles* mosquito host (left) to a human host (right). After the initial liver stage, the parasite begins its asexual intraerythrocytic cycle. Sexual forms, which develop from the intraerythrocytic parasites, can be transmitted to another mosquito. In the mosquito, parasites undergo meiotic and mitotic replication to form sporozoites, which can infect another human host.

Source: Bbkkk/Wikimedia Commons. Reproduced under the Creative Commons Attribution-Share Alike 4.0 International license <https://commons.wikimedia.org/wiki/File:Life_Cycle_of_the_Malaria_Parasite.svg>.

turn develop into sporozoites which make their way to the insect salivary glands ready to be injected into the next victim.

The degree of preference of the mosquito for human blood has important consequences for human malaria transmission. Mosquitos that take 10% of their blood meals on humans will transmit a malarial infection 100 times less frequently than mosquitoes that take all their blood meals on humans. In most parts of the world, the anthropophilic index or probability of a blood meal being on a human, is usually around 10–20% but in sub-Saharan Africa it is 80–100%. This probably is the most important factor responsible for the intensity of malarial transmission in tropical Africa today.

It was long believed that millions of years ago there was a common ancestor of *Plasmodium falciparum* that infected humans and chimpanzees and subsequently evolved separately. In contrast, *Plasmodium vivax* was thought to have arisen several hundred thousand years ago in south-east Asia following cross-species transmission from a macaque. However, the discovery of parasites closely related to *Plasmodium falciparum* and *Plasmodium vivax* in chimpanzees, bonobos, and western gorillas, all of which are African apes, has contradicted these theories. The current view is that both human pathogens emerged much more recently from species infecting African apes and genomic analysis supports this view.

Genomic analysis has shown that *Plasmodium falciparum* and *Plasmodium vivax* are only distantly related. At least seven *Plasmodium* species related to *Plasmodium falciparum* have been found in apes and the most closely related ones are found in wild gorillas. It now is believed that *Plasmodium falciparum* arose from zoonotic transmission of a gorilla parasite that has been given the name *Plasmodium prefalciparum* and molecular clock analysis suggests that this occurred 5,000–10,000 years ago. This time frame is consistent with the switch from a hunter–gatherer lifestyle to settled agricultural communities, the localized numbers of people being more favourable to mosquito transmission. Also, a number of mutations in the human genome (Box 21.1) can protect against malaria and the ones in glucose-6-phosphate dehydrogenase and β-globin are estimated to have arisen within the last 10,000 years. Extensive molecular epidemiological studies have shown that the ape relatives of *Plasmodium falciparum* do not appear to serve as a reservoir for human infection, indicating that cross-species events like the one that led to the emergence of *Plasmodium falciparum* as a human pathogen are very rare.

Box 21.1 NATURAL RESISTANCE TO MALARIA

Malaria has been a scourge of humans in many parts of the world for at least 5,000 years and because it can be a fatal infection, particularly to the young, there has been a strong selective pressure for individuals who naturally are resistant to the disease. A number of different malaria-protective mutations in humans have been identified and, given the role of red blood cells in human disease, it will come as no surprise that most of them are associated with erythrocyte structure and function.

Glucose-6-phosphate dehydrogenase (G6PD) deficiency is an X-linked recessive disorder and so most people who develop the disorder are male. There are a number of variants but the two major ones are the A variant that originated in west Africa and the D variant that originated in the eastern Mediterranean. One dietary consequence of G6PD deficiency is the risk of haemolytic anaemia from eating certain foods such as fava beans which are a staple in Mediterranean countries.

Thalassaemias are inherited blood disorders characterized by decreased haemoglobin production. There are two main types, alpha thalassaemia and beta thalassaemia. The severity of alpha and beta thalassaemia depends on how many of the four genes for alpha globin, or two genes for beta globin, are missing. Alpha thalassaemias often are found in people from south-east Asia, the Middle East, and China, while beta thalassaemia most often occurs in people of Mediterranean origin. Neither the thalassaemias nor G6PD deficiency is highly protective against death from malaria caused by *Plasmodium falciparum*: each reduces the risk by about a factor of two.

Much greater protection from malaria is provided by the sickle cell mutation which is a single point mutation in the gene for the beta chain of haemoglobin. Persons who are homozygous for this mutation have sickle cell disease and would rarely have survived to puberty before the twentieth century. Today, in developed countries they can live until they are forty or fifty years of age. Such a mutation could only have survived in the population if heterozygotes, those carrying one mutant and one normal gene, had an advantage great enough to balance the genetic cost of homozygotes. This advantage is a tenfold reduction in the risk of death from *Plasmodium falciparum* infection.

The Duffy antigen is a glycosylated protein located on the membrane of red blood cells which is an essential receptor for *Plasmodium vivax* merozoites. Most people who are indigenous to west and central Africa are homozygous for a mutation in the Duffy antigen gene that gives a high degree of protection from malaria due to *Plasmodium vivax*. There is evidence that the mutation arose more than once and another mutation with similar effect has been found in Papua New Guinea.

Large-scale testing of captive and wild apes has shown that many of them are infected with parasites that genetically are very close to human isolates of *Plasmodium vivax*. The current view is that *Plasmodium vivax* emerged from an ancient parasite that infected both apes and humans in Africa. In support of this is the observation that almost everyone in western and central Africa, but not Asia, is homozygous for the 'Duffy-negative phenotype' (Box 21.1). This fixation of the Duffy-negative mutation indicates that it is very old and it has driven the elimination of *Plasmodium vivax* from indigenous people in central and western Africa. The strains of *Plasmodium vivax* found in Asia represent a lineage that escaped out of Africa, possibly when humans first left the continent (~ 75,000 years ago) or perhaps more recently. Malaria caused by *Plasmodium vivax* occurs in Central and South America and probably got there as a result of the slave trade: yet another disease introduced there by colonial expansion. *Plasmodium simium* is found in monkeys in south-eastern Brazil and is believed to have arisen as a result of cross-species transmission of *Plasmodium vivax* from humans.

The presence of malaria in an area requires a combination of high human population density, high *Anopheles* mosquito population density, and high rates of transmission from humans to mosquitoes and from mosquitoes to humans. If any of these is lowered sufficiently, the parasite eventually disappears from that area, as happened in North America, Europe, and parts of the Middle East. An understanding of these interacting factors, along with improved medications, has led to a significant reduction in deaths due to malaria in the last thirty years.

The cheapest and most effective way to limit infectious diseases is to vaccinate the populations at risk. However, developing a vaccine against malaria has proved very challenging. The reason for this is that the parasite is relatively protected from attack by the body's immune system because for most of its human life cycle it resides within the liver and blood cells, making it relatively invisible to immune surveillance. However, circulating infected blood cells are destroyed in the spleen. To avoid this fate, *Plasmodium falciparum* displays adhesive proteins on the surface of the infected blood cells, causing the blood cells to stick to the walls of small blood vessels, thereby sequestering the parasite from passage through the general circulation and the spleen. A potential vaccine that targets the sporozoite stage of the parasite life cycle, in other words the stage it enters the body, is currently being trialled.

There are a number of medications that can help prevent or interrupt malaria in travellers to places where infection is common and these are mefloquine ('Lariam'), doxycycline, or the combination of atovaquone and proguanil ('Malarone'). Once a person gets the disease they are treated with antimalarial medications and the ones used depends on the type and severity of the disease. The malarial parasite has a limited ability to synthesize amino acids and obtains them by breaking down haemoglobin molecules in red blood cells. This results in the release of ferriprotoporphyrin IX which would be toxic to the parasite except that it gets polymerized to non-toxic hemozoin. Quinolines such as chloroquine, primaquine, and quinine block the polymerization step leading to lethal concentrations of the porphyrin. Quinolines are cheap, making them ideal for use in poorer countries but resistance to them now is widespread.

Box 21.2 THE VIETNAMESE WAR AND THE IDENTICATION OF ARTEMISININ AS AN ANTI-MALARIAL AGENT

The US Air Force dropped vast amounts of herbicides on North Vietnam during the Vietnamese war. The resultant deforestation led to a huge increase in malaria in bombed areas and the Viet Cong were losing many more men to malaria than from bullets. In desperation, the North Vietnamese leader Ho Chi Minh appealed to the Chinese for help. In response, Mao Zedong set up secret military project 523 to find suitable treatments from traditional Chinese remedies.

Artemisia annua (Figure 21.2) is a common herb found in many parts of the world, and has been used by Chinese herbalists for more than 2,000 years to treat fevers. In

1972, Tu Youyou isolated an effective anti-malarial compound from this plant and sub-sequently it was given the name artemisinin. In addition to artemisinin, Tu Youyou and her team developed a number of other compounds that can be used in combination with artemisinin, including lumefantrine, piperaquine, and pyronaridine.

In 2015, Youyou was awarded the Nobel Prize for Physiology or Medicine for her work on artemisinin, 113 years after Ronald Ross became the first British Nobel laureate for his work showing that malaria is transmitted by mosquitoes.

Figure 21.2. A specimen of *Artemisia annua*.

The most effective therapy for malaria is artemisinin (Box 21.2) or one of its derivatives, typically in combination with a longer-lasting partner drug in what is known as artemisinin-based combination therapy (ACT). The combinations are artemether/lumefantrine, artesunate/amodiaquine (ASAQ), artesunate/mefloquine, dihydroartemisinin/piperaquine, and artesunate/sulfadoxine-pyrimethamine. In each of these combinations, the artemisinin derivative rapidly kills the parasites, but is itself rapidly cleared from the body. The longer-lived partner drug kills the remaining parasites and provides some lingering protection from reinfection. Artemisinin contains an endoperoxide ring. This is cleaved inside red blood cells by the presence of haem to produce free radicals that in turn damage susceptible proteins thereby resulting in the death of the parasite. Unfortunately, inappropriate use of artemisinin has led to the development of resistance and there is nothing to replace it. Hopefully, genomics will help us to develop new anti-malarial drugs either through target identification or by revealing how to make an effective vaccine.

Key points

- There are five species of *Plasmodium* that can cause malaria in humans but 95% of cases are due to *Plasmodium falciparum* (Africa) and *Plasmodium vivax* (south-east Asia and South America). Malaria is transmitted by female mosquitos, usually *Anopheles* species, when they take a blood meal.

- The malarial parasite has a complex life cycle but when in humans it spends most of its time in liver and blood cells where it is protected from host defences. This makes the development of an effective vaccine difficult but a potential vaccine that targets the immediate post-infection phase is being trialled.

- Genomic analysis has shown that *Plasmodium falciparum* and *Plasmodium vivax* are only distantly related. *Plasmodium falciparum* arose following a species jump from gorillas to humans about 10,000 years ago when localized populations were developing following transition from hunter–gatherer to farmer.

- It was thought that *Plasmodium vivax* arose in Asia but the presence of close relatives in African apes plus the natural resistance of Africans to it suggest that it arose in Africa and 'escaped' to Asia.

- The most effective treatment for malaria is artemisinin-based combination therapy but resistance to artemisinin is developing.

Suggested Reading

Kariuki S.N. and Williams T.N. (2020) Human genetics and malaria resistance. *Human Genetics* **139**, 801–11.

Pronto W.R., Siegel S.V., Dankwa S., Liu W., Kemp A., et al. (2019) Adaptation of *Plasmodium falciparum* to humans involved loss of an ape-specific erythrocyte invasion ligand. *Nature Communications* **10**(1), 4512. doi:10.1038/s41467-019-12294-3

Sharp P.M., Plenderleith L.J., and Hahn B.H. (2020) Ape origins of human malaria. *Annual Review of Microbiology* **74**, 39–63. doi:10.1146/annurev-micro-020518-115628

Su X-Z., Lane K.D., Xia L., Sá J.M., and Wellems T.E. (2019) *Plasmodium* genomics and genetics: new insights into malaria pathogenesis, drug resistance, epidemiology and evolution. *Clinical Microbiology Reviews* **32**(4), e00019–19. doi:10.1128/CMR.00019-19

22

An Environmental Opportunistic Pathogen: *Cryptococcus*

Compared with bacteria and viruses, only a limited number of fungi are human pathogens and an even smaller number cause life-threatening disease. One fungus that is of major concern is the yeast *Cryptococcus*, particularly *Cryptococcus neoformans* and *Cryptococcus gatti*, that causes cryptococcosis. A surprising feature of these pathogens is their distribution in the environment. *Cryptococcus neoformans* was first isolated from peach juice in 1894 and since then has been found to be abundant in soil, particularly soil that is contaminated with the faeces of pigeons, chickens, turkeys, and other birds. *Cryptococcus neoformans* also has been isolated from the bark, hollows in the trunk, and decaying wood of a large number of trees. *Cryptococcus gatti* has not been found in bird droppings but, like *Cryptococcus neoformans*, has been isolated from a large number of trees. Furthermore, like *Legionella pneumophila* (p62), *Cryptococcus* can survive digestion by amoebae in the environment. As we shall see, many of the evolutionary adaptations of *Cryptococcus* that help it survive in these environments also help it to overcome human host defences.

The first description of cryptococcosis was made at the end of the nineteenth century but the incidence of the disease was low until the 1950s when there were increasing numbers of cases reported from west Africa. This increase probably was the result of the rise in the numbers of immunocompromised individuals because of the spread of HIV in the cities on the River Congo (p213). As the incidence of HIV/AIDS increased globally so too did cryptococcosis. Because of inequalities in treatment, 80% of HIV/AIDS-related cryptococcosis occurs in sub-Saharan Africa. In the last fifty years there has been an increase in non-HIV related cryptococcosis associated with the rise in organ transplants, where patients are given immunosuppressive drugs, and cancer chemotherapy.

Given the environmental niches in which *Cryptococcus* species are found, the most likely route of infection of humans is by inhalation of the yeasts

Microbiology of Infectious Disease. Sandy B. Primrose, Oxford University Press.
© Sandy B. Primrose (2022). DOI: 10.1093/oso/9780192863843.003.0022

or the spores that they form after mating. If the yeast is not cleared immediately by the immune system then there are a number of possible outcomes. The yeast may colonize the pulmonary system or lymph nodes where it can remain dormant for many years unless the host becomes immunocompromised when it will be reactivated. The extent of such colonization is apparent from the fact that approximately 70% of individuals have antibodies to *Cryptococcus* and cryptococcal granuloma in their lungs by adulthood. Alternatively, susceptible hosts may develop active pulmonary cryptococcosis, a form of pneumonia, or disseminated disease. In the latter case, any organ of the body can be affected but the brain is the preferred destination. Of the two pathogens, *Cryptococcus gatti* is more likely to cause pulmonary infections and *Cryptococcus neoformans* to cause brain infections. It is worth noting that cryptococcal infections are not restricted to immunocompromised individuals: about 20% of those with active disease are immunocompetent by standard criteria.

Cryptococcal yeast cells and spores are sufficiently small in size (4–7 microns) that, following inhalation, they are deposited deep in the respiratory system and end up in the alveolar space. Here the cryptococci encounter macrophages and what happens next determines the outcome for the patient. *Cryptococcus* species are surrounded by an extracellular polysaccharide capsule that is very effective at preventing phagocytosis and enabling it to cause disease as shown by the fact that acapsular strains are avirulent. If the host overcomes this capsular defence then the yeasts will be internalized by the macrophages and may be destroyed totally. However, like *Legionella pneumophila* (p62), *Cryptococcus* can survive and replicate inside the phagosomes. In this state they may form granulomas, aggregations of macrophages containing the cryptococci. Live cryptococci also can be expelled from macrophages by exocytosis and spread to other cells.

In order to cause brain infections, *Cryptococcus neoformans* needs to cross the blood–brain barrier which, under normal circumstances, prevents passage of microorganisms. *Cryptococcus* has found a number of ways of doing this but a key one is to use infected macrophages as a Trojan horse (Figure 22.1). Another way is for *Cryptococcus* to adhere to brain endothelial cells, be internalized, and then exit on the other side, a process known as transcytosis. The third mechanism is for *Cryptococcus* to damage the junctions between endothelial cells and squeeze through.

A number of factors that enable *Cryptococcus* to survive in the natural environment also act as virulence factors in humans and other animals. These include the polysaccharide capsule, the cell wall, melanin formation, and the secretion of enzymes. As noted earlier, the capsule is an essential virulence

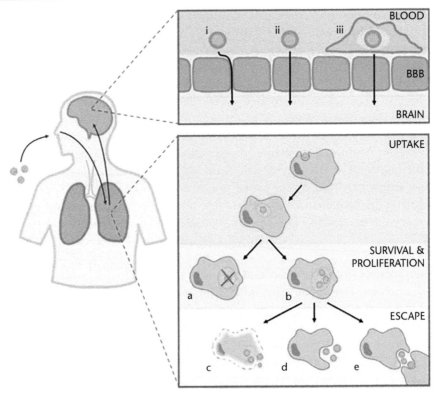

Figure 22.1. *Cryptococcus neoformans* interactions with host cells. Left, inhalation of infectious particles and dissemination to the brain. Right, events discussed in the text: Top, blood–brain barrier (BBB) traversal by *C. neoformans* may be paracellular (i), transcellular (ii), or via Trojan horse transit (iii). Bottom, possible fates of *C. neoformans* after engulfment include fungal clearance (a), proliferation (b), lytic escape (c), nonlytic exocytosis (d), and cell-to-cell transfer (e). Figure reproduced from Gaylord, Choy, and Doreing (2020) with permission.

factor. The cryptococcal cell can respond to different environmental conditions by changing the structure, size, and density of the capsule and this affects the ability of host macrophages to ingest it. Should *Cryptococcus* be ingested then the capsule protects it from reactive oxygen species and degradation. *Cryptococcus* also sheds capsular material causing the immune cells to rupture.

Melanin is a critical virulence factor for *Cryptococcus neoformans* as mutants lacking the pigment lose virulence. It is produced from diphenolic compounds by enzymes called laccases and in the central nervous system the melanin building blocks are neurotransmitters such as dopamine, norepinephrine, and epinephrine. Upon expression, laccases are loaded into secretory vesicles and deposited as spherical particles within the cell wall, using chitin as an anchoring molecule, or are secreted extracellularly. Melanin facilitates survival in the host by preventing oxidative damage and macrophage phagocytosis during infection. Melanin can also make cryptococcal cells resistant to the antifungal drug amphotericin B.

The biosynthesis of capsular material and melanin by *Cryptococcus* requires copper and iron. *Cryptococcus neoformans* has genes for copper sensing and copper transport and high expression of these genes has been seen in isolates from macrophages and the brain. It also has high affinity iron transporters and a siderophore transporter. *Cryptococcus neoformans* does not produce siderophores (p19) but can acquire them if they are expressed by other microbes in its vicinity, illustrating how co-infection can increase pathogenicity.

Cryptococci secrete a number of exoenzymes such as proteases, phospholipases, and urease. The best studied are the phospholipases. These cause destabilization of membranes and enable the fungus to penetrate lung tissue as well as promoting dissemination in blood and lymph. One in particular, phospholipase B, enhances cryptococcal survival in macrophages by enabling it to take up macrophage arachidonic acid and convert it to eicosanoids that suppress the host immune response. The function of cryptococcal proteases in the infection process is not known but mutants lacking them have decreased virulence. Urease hydrolyses urea to ammonia and carbonate which will increase the local pH and damage membranes. Mutant analysis shows that urease is important for dissemination to the brain but the mechanism is unclear.

A novel virulence factor in *Cryptococcus* is cell size. In the wild, *Cryptococcus* strains are haploid with genomes of size from 16–19 Mb and contain 14 chromosomes. Their normal cell size is 4–7 microns but in the lungs 20–30% of them convert to Titan cells that are polyploid and anywhere in size from 10–100 microns. A trigger factor for the formation of Titan cells is peptidoglycan from the normal bacterial flora of the lungs. Key features of Titan cells are a cell wall that is ten times thicker than that seen in normal cells and a much denser, highly cross-linked capsule. These modifications reduce phagocytosis and promote dissemination to the central nervous system. Titan cells have at least four copies of the genome and can have sixteen or more copies. This polyploidy results in resistance to fluconazole, a key drug in the treatment of cryptococcal infections (Box 22.1), by increasing the number of copies of the target gene (*ERG11*) and the genes for fluconazole efflux pumps. Titan cells can undergo genome reduction and produce daughter cells that are haploid. However, in the presence of fluconazole, these daughter cells will be aneuploid and have multiple copies of chromosomes encoding azole resistance. In addition to forming Titan cells, *Cryptococcus* can form micro cells that are less than one micron in size and which are believed to be more effective in dissemination to the brain.

Cryptococcus is not spread from human to human nor from humans to the environment: it is an opportunistic pathogen whose natural home is the environment. Consequently, traits that increase pathogenicity must arise in the environment and not by evolution in humans. For example, many strains of *Cryptococcus neoformans* have multiple genes for inositol transport and metabolism and this is an adaptation for growth on trees. Inositol is abundant in the human brain and is required for cryptococcal virulence and so growth on trees can facilitate meningitis caused by *Cryptococcus neoformans*. Another example is melanin which protects the yeast from the damaging effects of ultraviolet light and solar radiation, a key requirement for an organism exposed to the African sun. The amount of exposure to the sun will influence the amount of melanin in a particular isolate and hence its resistance to phagocytosis should it infect a human. Another environmental effect, but one driven by agricultural practice, is an increase in azole resistance in clinical isolates driven by the use of azoles as agricultural fungicides.

Box 22.1 ANTIFUNGAL DRUGS

Azole antifungal agents, which can be either imidazoles or triazoles, prevent the synthesis of ergosterol, a major component of fungal plasma membranes by inhibiting the cytochrome P-450-dependent enzyme lanosterol demethylase. This enzyme also plays an important role in cholesterol synthesis in mammals. However, when azoles are present in therapeutic concentrations, their antifungal efficacy is attributed to their greater affinity for the fungal enzyme than for the mammalian enzyme. Fluconazole is the azole used to treat cryptococcosis.

The polyene compounds are so named because of the alternating conjugated double bonds that constitute a part of their structure. These drugs interact with sterols in cell membranes (ergosterol in fungal cells; cholesterol in human cells) to form channels through the membrane, causing the cells to become leaky. Because of their action on mammalian as well as fungal membranes they have considerable toxicity. The one used to treat *Cryptococcus* infections is amphotericin B.

Like the azole antifungals, allylamines interfere with an enzyme that is involved in the creation of the fungal cell membrane. The most widely used allylamine is terbinafine, but it is mostly used topically, and thus not for cryptococcosis, because it has many side effects when taken orally.

Echinocandins are to fungi what penicillins are to bacteria as they inhibit the formation of the fungal cell wall. They are well tolerated but display negligible activity against *Cryptococcus*.

Genetic analysis has shown that *Cryptococcus neoformans* can be divided into three distinct lineages. The VNI and VNII lineages are highly clonal and globally distributed whereas the third lineage, VNB, is genetically diverse yet largely restricted to sub-Saharan Africa. The VNB lineage is mostly isolated from mopane trees that are common in the African savannah, which would explain its geographical restriction. However, cryptococcal infections were not seen outside Africa until relatively recently so how did they become global in the absence of human–human transmission? The VNI lineage is associated with trees and pigeon droppings and so global spread could be associated with bird migrations. However, VNII isolates have only been made clinically so what is its reservoir and how has it spread?

Key points

- *Cryptococcus neoformans* and *Cryptococcus gatti* are yeasts that can cause severe, life-threatening infections of the lungs and brain, particularly in immunocompromised individuals.
- Cryptococci are opportunistic pathogens. Their natural habitat is the environment where they have evolved a series of mechanisms to resist radiation, desiccation, high temperatures, and digestion by protists such as amoebae.
- Infections of humans occurs following inhalation of the yeasts or their spores. The factors that enable survival in the environment enable the yeasts to escape many of the defences of the human immune system.
- Because human-to-human transmission does not occur, the genetic traits of any clinical isolate must have been selected in the environment.
- Cryptococcal disease initially was restricted to sub-Saharan Africa but now has spread to other parts of the world. It is not clear how this has occurred given that there is no human-to-human transmission.

Suggested Reading

Desjardins C.A., Giamberardino C., Sykes S.M., Yu C-H., Tenor J.L., et al. (2017) Population genomics and the evolution of virulence in the fungal pathogen *Cryptococcus neoformans*. *Genome Research* **27**, 1207–19.

Gaylord E.A., Choy H.L., and Doering T.L. (2020) Dangerous liaisons: interactions of *Cryptococcus neoformans* with host phagocytes. *Pathogens* **9**(11), 891. doi:10.3390/pathogens9110891

Iyer K.R., Revie N.M., Fu C., Robbins N., and Cowen L.E. (2021) Treatment strategies for cryptococcal infection: challenges, advances and future outlook. *Nature Reviews Microbiology* **19**, 454–66. doi:10.1038/s41579-021-00511-0

Montoya M.C., Magwene P.M., and Perfect J.R. (2021) Associations between *Cryptococcus* genotypes, phenotypes, and clinical parameters of human disease: a review. *Journal of Fungi* 7(4), 260. doi:10.3390/jof7040260

Sabiiti W. and May R.C. (2012) Mechanisms of infection by the human fungal pathogen *Cryptococcus neoformans*. *Future Microbiology* **7**, 1297–313.

Zafar H., Altamirano D.S., Ballou E.R., and Nielsen K. (2019) A titanic drug resistance threat in *Cryptococcus neoformans*. *Current Opinion in Microbiology* **52**, 158–64.

23

The Most Famous Plant Pathogen:
Phytophthora infestans

Although potatoes were introduced to Europe from Central America in the late sixteenth century, it was nearly 200 years before they gained popularity. The driving force for adoption was famine plus encouragement from such notable figures as Catherine the Great of Russia, Frederick the Great of Prussia, and King Louis XVI of France. Agronomically, potatoes are easy to sow and to harvest and require no processing. They also have a threefold higher yield per acre than wheat. Nutritionally, they are high in calories and fibre but low in fat and have no cholesterol. They also are very rich in B vitamins, vitamin C, and minerals as well as containing vitamin A. What potatoes are missing is fat, protein, and vitamin D but these can be provided by milk.

Much of the land in Ireland is not suitable for growing arable crops such as wheat. In the west of the country, the land is very poor but will support the growth of potatoes. One acre of potatoes plus a cow for milk could support a family for a year. Initially there was enough land available for rent but as the population increased, plots of land were divided and then sub-divided until they were barely large enough to sustain a family. This led to the widespread adoption of a single variety of potato: the Lumper. It is pretty tasteless but a very heavy cropper with minimum attention needed for cultivation. In time, it became virtually the only potato variety being grown in Ireland.

Late potato blight was first recorded in the United States in 1843 in Philadelphia and New York City and by 1845 it was found from Illinois to Nova Scotia, and from Virginia to Ontario. In 1845 it also crossed the Atlantic Ocean. All of the potato-growing countries in Europe were affected, but the potato blight hit Ireland the hardest because of Ireland's near dependence on a single variety of potato. The organism causing late potato blight is an oomycete (Box 23.1) called *Phytophthora infestans*. It spreads best when conditions are warm and humid and the weather in Ireland in 1845 was ideal. Initially, white mould would have appeared on the underside of leaves followed by a blackening and collapse of the whole plant (Figure 23.1). Infected

Microbiology of Infectious Disease. Sandy B. Primrose, Oxford University Press.
© Sandy B. Primrose (2022). DOI: 10.1093/oso/9780192863843.003.0023

tubers (Figure 23.2) would have quickly decayed as a result of infestation of secondary soft bacterial rots. Seemingly healthy tubers would have rotted later when in store. Commentators at the time noted that you could tell an infected crop by the smell of rotting surrounding the field.

Figure 23.1. A potato field heavily infected by *Phytophthora infestans*.

Source: Björn Andersson (SLU)/Wikimedia Commons. Reproduced under Creative Commons Attribution-Share Alike 4.0 International license <https://commons.wikimedia.org/wiki/File:Potato_field_heavily_infected_by_P._infestans.JPG>.

Figure 23.2. A potato tuber infected with *Phytophthora infestans*. <https://en.wikipedia.org/wiki/Phytophthora_infestans#/media/File:Phytophtora_infestans-effects.jpg>.

In 1845, half of the Irish potato crop was destroyed and the following year all of the crop was lost. We now know that the pathogen is a hemibiotroph: an organism that is parasitic in living tissue for some time and then continues to live in dead tissue. In Ireland, it would have remained viable within rotting tubers that were left in the ground or in cull piles, as they would have been by the starving people. These infected tubers would have acted as an inoculum at the start of the next growing season. Although the harvest improved in 1847, the blight returned with full force the following two years. By 1850, approximately 1 million people had died and another million had emigrated to North America. By 1900, the population of Ireland had halved to 4 million people.

Once the reproductive cycle of *Phytophthora infestans* was deciphered it became very clear why it is such a devastating pathogen, not only of potatoes but tomatoes as well. During asexual reproduction, sporangia are produced on sporangiophores that grow from infected tissue. These sporangia can germinate directly and develop mycelial growth that can infect leaves, stems, and fruit. At lower temperatures, sporangia develop into motile zoospores that then will germinate and cause infections. Initially there will be no obvious symptoms on infected plants but after two days there will be small areas of necrosis (Figure 23.3). Within another two days, in the presence of free water or high humidity, up to 300,000 sporangia will be produced per necrotic lesion. Each of these sporangia can go on to infect more plants. This explains the difficulty in managing outbreaks: whole fields can go from being slightly diseased to completely destroyed within a few days! To make matters worse, an RNA virus (PiRV-2) has been found in some strains of *Phytophthora infestans* that stimulates sporangia production.

Box 23.1 OOMYCETES

The oomycetes or 'water moulds' are a group of several hundred organisms that include some of the most devastating plant pathogens. The diseases they cause include late blight of potato (*Phytophthora infestans*), downy mildew of grape vine (*Plasmopara viticola*), and sudden oak death (*Phytophthora ramorum*). For many years, oomycetes were thought to be fungi because of their filamentous growth and reproduction via spores. Based on molecular phylogenetics it now is known that they are more closely related to algae and green plants than to fungi. One of the most distinguishing characteristics of oomycetes is the production of zoospores from sporangia. Zoospores can swim in water films on leaf surfaces and in soil water and so their spread is facilitated by humid or damp conditions. Sexual reproduction also can occur in *Phytophthora infestans*, resulting in the production of oospores, but requires the presence of A1 and A2 mating types. Oospores of many oomycetes have been shown to be able to survive for years in soil, making their elimination difficult without chemical intervention.

Figure 23.3. Early symptom of late blight (*Phytophthora infestans*) on the underside of a potato leaf.

Source: Agricultural Research Service, the United States Department of Agriculture (Public domain) <https://commons.wikimedia.org/wiki/File:Phytophtora_infestans-effects.jpg>.

Today, potatoes are the third largest food crop in the world and late blight continues to wreak havoc. The centre of diversity for potatoes is in the northern Andes of South America. There, each farmer grows tens of different varieties and there are over 200 varieties in cultivation. Different varieties have differing susceptibility to *Phytophthora infestans* and so late potato blight never is a major problem. By contrast, in North America and Europe, establishment of late blight is favoured by the growth of a small number of varieties and sometimes even a single variety (monoculture) to meet the needs of major food corporations. This has led to widespread use of chemicals and two groups of compounds are used routinely: topical protectants and systemic fungicides. Phenylamides are in the latter category and are widely used to prevent plant infections with oomycetes with twelve to fifteen applications per year! Despite so many applications, or perhaps because of it, resistance arises readily by mutation, and resistant clones are widespread.

A key part of managing plant disease caused by *Phytophthora infestans* is understanding what strains of it are in circulation. This has been facilitated by the availability of rapid genome sequencing and has allowed plant pathologists to determine the evolutionary relationship between strains from nineteenth-century outbreaks, such as the Irish potato famine, and strains from more recent outbreaks. As *Phytophthora infestans* is a eukaryotic organism it has two genomes that can be sequenced: the large nuclear genome and the much smaller mitochondrial genome. Analysis of nineteenth-century herbarium specimens from North America and Europe showed that they all belonged to a single lineage: the HERB-1 mitochondrial DNA haplotype and the FAM-1 nuclear genotype. There was little genetic diversity between isolates from different herbaria and this lineage persisted for over fifty years. Thus, the Irish outbreak was just part of a much wider geographic late blight pandemic. Until the early 1990s, a single lineage (US-1) dominated the global population and it was thought that it was identical to FAM-1. However, genomic analysis has shown that US-1 is related to, but distinct from, FAM-1 and the two lineages are believed to have separated in the early nineteenth century. Since the 1990s the situation has become more complex. Multiple lineages exist that vary between countries and distinct clonal lineages are constantly replacing older ones. In most cases, the pathogen is reproducing asexually because isolates of it belong to a single mating type. However, the pathogen appears to be sexually reproducing in Mexico, the Netherlands, and Northern Europe because both mating types are found in pathogen isolates from these countries and in a 1:1 ratio.

In plants, susceptibility or resistance to disease is governed by the 'gene-for-gene' relationship. Inheritance of both resistance in the host and the pathogen's ability to cause disease is controlled by pairs of matching genes. One is a plant gene called the resistance (*R*) gene. The other is a pathogen gene called the avirulence (*Avr*) gene. Plants producing a specific *R* gene product are resistant towards a pathogen that produces the corresponding *Avr* gene product (Figure 23.4). According to the zigzag model of plant–pathogen interaction, there are two phases of plant defence. In the first phase, the plant recognizes conserved molecules or structures on the pathogen (pathogen associated molecular patterns or PAMPs) and this triggers the plants defence responses (pathogen triggered immunity). To circumvent this, pathogens deliver effector proteins inside host cells. In the second phase of defence,

Plant genotype / Pathogen genotype	RR or Rr	rr
AA or Aa	Resistance	Disease
aa	Disease	Disease

Figure 23.4. The gene for gene relationship between a plant pathogen and its host. A and a are different alleles of a particular *Avr* gene.

the plants perceive the effectors through the resistance (R) genes and activate a more robust and faster defence response known as effector-triggered immunity (ETI). When one or more pathogen effectors neutralize the first line of defence, the pathogens successfully infect susceptible hosts. In the absence of effective R proteins, ETI is overcome, eventually leading to effector-triggered susceptibility. Multiple shifts between susceptibility and immunity occur because of coevolution of effectors in pathogens and the corresponding R genes in plant hosts.

Gene-for-gene relationships are important in the context of breeding resistance to plant pathogens. In the case of *Phytophthora infestans*, one group of *Avr* gene products are effectors known as RxLR proteins. These are secreted proteins that contain an arginine-X-leucine-arginine (where x can be any amino acid) sequence at the amino terminus of the protein that is required for translocation into host cells. How this RxLR motif facilitates entry is not known but it embodies the complete machinery that the pathogen needs to deliver effectors into host cells. This is in contrast with the bacterial type III secretion system, used by pathogens such as *Pseudomonas syringae* (p87), which requires a multitude of proteins to accomplish the same task.

In contrast to their N-termini, the C-terminal domains of RxLR proteins show extensive sequence divergence and contain determinants for recognition by specific intracellular NB-LRR resistance (R) proteins. The name of these proteins derives from the fact that they have a central nucleotide-binding (NB) domain and a leucine-rich repeat (LRR) domain. In the presence of the corresponding R protein, the RxLR effector is recognized as an avirulence (*AVR*) factor resulting in a hypersensitive response that blocks further growth of the pathogen. In the absence of appropriate R proteins, RxLR effectors promote virulence of the pathogen by suppressing defence responses.

The strategy for breeding late blight-resistant potatoes during the first half of the twentieth century was the introduction of dominant R genes from the Mexican wild potato species *Solanum demissum* which intrinsically is resistant to *Phytophthora infestans*. However, the hybrids failed to provide durable resistance owing to the evolution of new races of the pathogen. Other R genes were bred in from other *Solanum* species but, again, resistance was short lived. At the present time, no single R gene has been effective for sufficiently long to have contributed much to late potato blight suppression in production agriculture. Genome sequencing has provided an explanation for this failure.

The genome of *Phytophthora infestans* is considerably larger (240 Mbp) than that of most other *Phytophthora* species whose genomes have been sequenced: *Phytophthora sojae* (stem and root rot of soybean) has a 95 Mbp genome and *Phytophthora ramorum* (sudden oak death) has a 65 Mbp genome. The late blight genome is unusual in having two compartments: gene-dense regions containing predominantly housekeeping genes and gene-sparse regions enriched for repeat sequences and gene families encoding effector proteins that are involved in pathogenicity. The gene-poor, repeat-rich loci are dynamic regions of the genome that facilitate evolution of *Phytophthora* species by promoting genome plasticity and enhancing genetic variation of effector genes.

A key difference between *Phytophthora infestans* on the one hand, and *Phytopthora sojae* and *Phytophthora ramonum* on the other, is a much higher number of effector genes in the former. For example, *Phytophthora infestans* has nearly 200 genes for the Crinkler (CRN for CRinkling and Necrosis) protein family. The exact function of the Crinkler proteins is not known but they cause a characteristic leaf crinkling of infected plants. More important, 560 genes for RxLR proteins have been found in *Phytophthora infestans* and only 16 of these are found in the other 2 species. Given this redundancy of RxLR genes, it is not surprising that the pathogen can overcome newly developed resistant varieties of potato.

One strategy for developing resistant varieties is to use 'gene stacking' where multiple *R* genes are introduced using genetic engineering techniques to avoid the need for lengthy plant breeding. Cultivars with four or more *R* genes have been produced and have remained blight free for several years. However, experience with the cultivar Pentland Dell suggests that gene stacking will fail. This cultivar contains genes *R1*, *R2*, and *R3* and was resistant to the strain of *Phytophthora infestans* that was in circulation at the time of introduction (1963). Four years after Pentland Dell was introduced there were serious outbreaks of blight in it: a new strain of the pathogen had been selected. Worse still, some of these strains were resistant to new *R* genes that had not yet been deployed in potato cultivars. It is with good reason that one plant pathologist has dubbed *Phytophthora infestans* as the '*R* gene destroyer'!

Key points

- Late blight of potatoes and other solanaceous plants is caused by the oomycete *Phytophthora infestans*. It reproduces from sporangia and up to 300,000 sporangia can be produced from a single necrotic lesion meaning that disease spreads quickly. The organism also can grow in dead tissue such as diseased tubers.

- The best defence against crop destruction is to grow a wide variety of different potatoes and the Irish Famine was caused by dependence on a single variety. Global food corporations are driving a dependence on just a few varieties and these can be maintained only by heavy fungicide use.

- Attempts have been made to breed resistant potatoes by introducing resistance determinants from other solanaceous crops but the pathogen quickly overcomes the new varieties. Genome sequencing has provided an explanation: the genome is enriched for many gene families of effectors.

Suggested Reading

Cai G., Fry W.E., and Hillman B.I. (2019) PiRV-2 stimulates sporulation in *Phytophthora infestans*. *Virus Research* **271**, 197674. doi:10.1016/j.virusres. 2019.197674

Fry W. (2008) *Phytophthora infestans*: the plant (and *R* gene) destroyer. *Molecular Plant Pathology* **9**, 385–402.

Goss E.M., Tabima J.F., Cooke D.E.L., Restrepo S., Fry W.E., et al. (2014) The Irish potato famine pathogen *Phytophthora infestans* originated in central Mexico rather than the Andes. *Proceedings of the National Academy of Sciences* **24**, 8791–6.

Jo K-R., Kim C-J., Kim S-J., Kim T-Y., Bergervoet M., et al. (2014) Development of late blight resistant potatoes by cisgene stacking. BMC Biotechnology **4**, 50. doi:10.1186/1472-6750-14-50

Raffaele S., Win J., Cano L.M., and Kamoun S. (2010) Analyses of genome architecture and gene expression reveal novel candidate virulence factors in the secretome of *Phytophthora infestans*. *BMC Genomics* **11**, 637. doi:10.1186/1471-2164-11-637

Saville A.C., Martin M.D., and Ristaino J.B. (2016) Historic late blight outbreaks caused by a widespread dominant lineage of *Phytophthora infestans* (Mont.) de Bary (2016) *PLoS One* **11**(12), e0168361. doi:10.1371/journal. pone.0168381

Yoshida K., Schuenemann V.J., Cano L.M., Pais M., Mishra B., et al. (2013) The rise and fall of the *Phytophthora infestans* lineage that triggered the Irish potato famine. *eLife* **2**, e00731. doi:10.7554/eLife.00731

Part IV
Viral Pathogens

24

A Virus That Promotes Its Own Transfer: Tomato Spotted Wilt Virus

Plant viruses are responsible for the loss of crops worth billions of dollars every year and most of these viruses are transmitted by insects such as aphids, thrips, leafhoppers, planthoppers, and whiteflies. The viruses are transmitted during insect feeding and the host range of plant viruses is largely governed by the plant preferences of the insect vector. The relationship between the virus and its insect vector can be complex. There are three different modes of transmission: non-persistent, semi-persistent, and persistent. In non-persistent transmission, the insect acquires the virus on its mouthparts when feeding on an infected plant and contaminates the next plant that it feeds on if it is uninfected, that is, transmission essentially is mechanical. Semi-persistent viruses enter the gut but do not penetrate insect tissues. Persistent viruses penetrate the insect gut wall and make their way to the salivary glands prior to being injected into the next host.

In an earlier chapter (p143), we saw how phytoplasmas can modify its plant and insect hosts and, from an evolutionary point of view, it would make sense for plant viruses to do the same. However, plant viruses have a very much smaller genome than phytoplasmas but despite this, they can alter insect behaviour to enhance their spread. For example, aphids that have acquired Barley yellow dwarf virus (BYDV) prefer to feed on uninfected wheat plants while aphids that are virus-free prefer to feed on plants infected with BYDV. At the same time, virus infection alters the secondary metabolism of the plant such that there is a change in the spectrum of volatile organic compounds produced. This in turn can affect the feeding behaviour of insect pests. The question is, how do plant viruses with their small genomes achieve this? An answer will be provided by an analysis of Tomato spotted wilt virus (TSWV) which belongs to a group of viruses known as Bunyaviruses (Box 24.1).

TSWV comes second to tobacco mosaic virus in terms of the damage caused to the global agricultural industry because it has a very wide host

Microbiology of Infectious Disease. Sandy B. Primrose, Oxford University Press.
© Sandy B. Primrose (2022). DOI: 10.1093/oso/9780192863843.003.0024

range. It can infect over 1,500 plant species distributed across more than 90 plant families including many used as food, fibre, and ornamental crops (Figure 24.2). The vectors of TSWV are thrips, minute (< 1 mm long) insects with fringed wings that feed by the 'punch and suck' method. Briefly, the mandible is used to cut into the food plant; saliva is injected and the maxillary stylets, which form a tube, are then inserted, and the semi-digested food pumped from ruptured cells. This process takes less than five minutes and leaves cells destroyed or collapsed, and a distinctive silvery or bronze scarring on the surfaces of the stems or leaves where the thrips have fed. At least ten different species of thrips transmit TSWV but the one that predominates globally is the western flower thrips (*Frankliniella occidentalis*, Figure 24.3). Because it can feed on hundreds of plant species, the virus is widely disseminated and in a single season a particular viral lineage could pass through multiple crop species. Although many crops are grown as annuals there are enough perennial plants that can act as winter reservoirs for the virus.

The relationship between TSWV and its vector is unique in that adult thrips are only able to transmit TSWV if acquisition occurs in the first instar and early second larval thrips stages. After the larvae feed on infected plants, the virus spreads from the insect midgut through the pupal stages and accumulates in the salivary glands of adults where major viral replication occurs. That is, TSWV is a persistent plant virus as defined earlier. However, when thrips feed as adults on virus-infected plants, the virus remains restricted to the midgut, and adults cannot transmit the virus because the midgut acts as a barrier to virus escape during certain developmental stages. Both TSWV

Box 24.1 BUNYAVIRUSES

Bunyaviruses are a large group of spherical, membrane-bound viruses that cause serious human, veterinary, and plant diseases. These include Crimean Congo haemorrhagic fever and Lassa fever in humans, Rift Valley fever in humans and animals, and TSWV in plants. The viruses have a characteristic RNA genome structure: it is divided into three segments designated small (S), medium (M), and large (L). In TSWV the segment sizes are 2.9 Kb, 5.4 Kb, and 8.9 Kb respectively. The RNA segments of Bunyaviruses are single-stranded and exhibit a pseudo-circular structure due to each segment's complementary ends (Figure 24.1). The L segment encodes the RNA-dependent RNA polymerase, necessary for viral RNA replication and mRNA synthesis. The M segment encodes the viral glycoproteins (Gc and Gn), which project from the viral surface and aid the virus in attaching to and entering the host cell. The S segment encodes the nucleocapsid protein (N). The genome also encodes two other proteins, NS and NSm, that are not structural parts of the virus. In the case of TSWV, these proteins are involved in overcoming plant host defences (NSs) and in spread of the virus through infected plants (NSm).

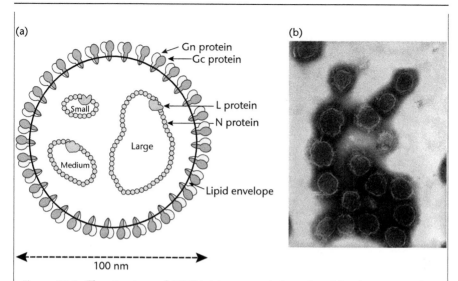

Figure 24.1. The structure of TSWV. (a) representation of a virion in cross-section, (b) negative-stained transmission electron microscopy photograph.

Source: Wikimedia Commons (composite of two Public domain elements). Reproduced under Creative Commons Attribution-Share Alike 4.0 International license. https://commons.wikimedia. org/wiki/File:Peribunyavirus_virion_structure.gif.

envelope glycoproteins (Gn and Gc) are critical for thrips infection. They play an essential role in attachment of the virus to the thrips gut and subsequent infection of its cells.

When an adult thrips feeds on an uninfected plant, the virus is released to the wounded plant tissue, directly into the cell's cytoplasm. As the virus gets uncoated its single-stranded RNA gets converted to a double-stranded form by the virally encoded RNA-dependent RNA polymerase. However, plants have a defence mechanism. Double-stranded RNA is recognized by a plant protein, the Dicer-like protein, and cleaved to produce molecules about twenty-one base pairs in length known as short interfering RNA (siRNA). These siRNA molecules are used by the plant's RNA-induced Silencing Complex to target viral messenger RNA and stop translation. In response, the viral NSs protein binds the siRNA molecules and stops them targeting viral RNA. The NSs protein functions in an identical way in insects which also have a Silencing Complex. As the virus successfully replicates in plant cells, it spreads to the entire plant with the aid of its movement protein (NSm), using the plasmodesmata for short distances and the phloem–xylem system to spread to the rest of the plant tissue.

Figure 24.2. Plants infected with TSWV. An infected tomato (top) and an infected sweet pepper plant (bottom).

Source: Downtowngal/Wikimedia Commons (top), reproduced under Creative Commons CC0 1.0 Universal Public Domain Dedication. https://commons.wikimedia.org/wiki/File:Tomato_with_Tomato_Spotted_Wilt_Virus.jpg. Carlos Gonzalez/Wikimedia Commons (bottom), reproduced under Creative Commons Attribution-Share Alike 3.0 Unported license. https://commons.wikimedia.org/wiki/File:Pepper_infected_with_tomato_spotted_wilt_virus.jpg.

When a plant is wounded, as it is by thrips when they feed, it responds by producing jasmonic acid and its methyl ester. These jasmonates, in turn, induce the plant to produce volatile organic compounds (VOCs) with acrid smells, such as terpenes, that act as repellents to herbivores. It has been shown that TSWV-infected plants produce significantly less terpenes than uninfected plants. This happens because the TSWV NSs protein binds to

Figure 24.3. The western flower thrips (*Frankliniella occidentalis*) https://commons. wikimedia.org/wiki/File:Frankliniella_occidentalis_5364132-LGPT.jpg.

the transcription factors that regulate the jasmonate-signalling pathway and suppress jasmonate formation. But the virus does more than this because it stimulates the synthesis of salicylic acid, a plant hormone involved in protection from bacterial and fungal pathogens. It will be recalled from the discussion on *Pseudomonas syringae* (p87) that the salicylic acid and jasmonate pathways are mutually antagonistic and so the increase in salicylic acid could further repress the expression of genes controlled by jasmonates. The net effect is that TSWV plants produce less terpene than uninfected plants, making them more attractive to thrips and thus facilitating the spread of the virus. As well as suppressing terpene formation, infection with TSWV results in a doubling of the free amino acid content of plants. This not only facilitates egg formation by the thrips but growth of virus-infected thrips populations as well.

TSWV can replicate in at least ten species of thrips and infect over 1,000 plant species and this raises the question as to how much the virus genome needs to mutate to enable the virus to adapt to so many hosts. This is an important question because, in other RNA viruses, mutations that increase fitness in one host often decrease fitness in another. Mutants of TSWV arise readily because RNA replication is error-prone and genome sequencing has revealed the selection pressures imposed by different hosts and how the virus adapts to them. When the virus is cycled between two unrelated plant hosts, the proteins that undergo change are NSs and NSm. This is not surprising given the role of these proteins in overcoming plant defences and spreading from cell to cell. Several amino acid variants are found to be enriched in plants versus thrips and these were predominantly in NSs. This is not surprising given that NSs is involved in suppressing RNA silencing in both thrips and plants.

Because TSWV is transmitted persistently by thrips, control of the disease has focussed on managing the levels of thrips in horticultural settings. A number of insecticides are effective against thrips but the insects can rapidly develop insecticide resistance if the insecticides are over-used or not changed regularly. Thrips also can evade the application of insecticides by hiding on the underside of leaves. Attempts to breed virus-resistant plants using both conventional breeding and genetic manipulation have not resulted in long-term success and currently TSWV is the undoubted winner in the arms race.

Key points

- TSWV is a persistent plant virus; that is, one that replicates in both its insect vector (thrips) and plants.
- The viral NSs protein interferes with the plant's RNA-induced silencing complex that is the first line of defence against virus infection.
- The NSs protein also interferes with the plant's jasmonate-signalling pathway and suppresses the formation of insect-repellent terpenes. This makes infected plants more attractive to thrips than uninfected plants. Infected plants also have a higher free amino acid content which favours thrips reproduction.
- TSWV adapts to new plant hosts by the selection of mutants in its NSs and NSm proteins, mutation being facilitated by the error-prone nature of RNA replication.

Suggested Reading

Nachappa P., Challacombe J., Margolies D.C., Nechols J.R., Whitfield A.E., and Rotenberg D. (2020) Tomato spotted wilt virus benefits its thrips vector by modulating metabolic and plant defense pathways in tomato. *Frontiers in Plant Science* **11**, 575564. doi: 10.3389/fpls.2020.575564

Nilon A., Robinson K., Pappu H.R., and Mitter N (2021) Current status and potential of RNA interference for the management of Tomato spotted wilt virus and thrips vectors. *Pathogens* **10**(3), 320. doi:10.3390/pathogens10030320

Ruark-Seward C.L., Bonville B., Kennedy G., and Rasmussen D.A. (2020) Evolutionary dynamics of Tomato spotted wilt virus within and between alternate plant hosts and thrips. *Scientific Reports* **10**(1), 15797. doi:10.1038/s41598-020-72691-3

Wu X., Xu S., Zhao P., Zhang X., Yao X., et al. (2019) The Orthotospovirus nonstructural protein NSs suppresses plant MYC-regulated jasmonate signaling leading to enhanced vector attraction and performance. *PLOS Pathogens* **15**(6), e1007897. doi:10.1371/journal.ppat.1007897

25

Morbilliviruses: Measles, Rinderpest, and Canine Distemper

Morbilliviruses are a group of closely related RNA viruses (Box 25.1) that cause serious diseases that have significant human health, economic, and conservation implications. The group includes measles virus that infects primates, rinderpest that infected cattle before it was eradicated globally, peste des petits ruminants virus (PPRV) which causes disease in small ruminants (sheep, goats), canine distemper virus which infects dogs and many other carnivores (lions, etc.), phocine distemper virus in seals, and cetacean morbillivirus that infects dolphins and whales. The viruses are highly infectious, spread via the respiratory route, cause profound immune suppression, and can cause outbreaks associated with high morbidity and mortality.

Box 25.1 COMMON FEATURES OF MORBILLIVIRUSES

All the viruses have a single-stranded RNA genome ranging in size from 15,690 to 16,050 nucleotides and the genome size always can be divided by 6. There are six genes in the order:

  N nucleocapsid
  P/V/C phosphoprotein gene encoding the non-structural V and C proteins
  M matrix protein
  F fusion glycoprotein
  H haemagglutinin
  L large protein, which is the enzyme RNA-dependent RNA polymerase

There is a large, untranslated region between the genes for the M and F proteins. The non-structural V and C proteins are innate immunity antagonists that help the virus to escape host immune response.

  The process of infection begins when the H protein binds to signalling lymphocyte activation molecules (SLAM) on the surface of immune cells. This induces conformational change in the viral F protein which fuses the viral membrane with the

---

Microbiology of Infectious Disease. Sandy B. Primrose, Oxford University Press.
© Sandy B. Primrose (2022). DOI: 10.1093/oso/9780192863843.003.0025

plasma membrane of target immune cells. After virus multiplication, infected immune cells fuse with epithelial cells via the nectin 4 protein on the basal side of the epithelial cells. The virus is transported through the epithelium and released to the outside.

Infection with measles virus confers lifelong immunity and, because there is no animal reservoir, measles is principally a disease of childhood. Maintenance of measles virus in a population requires a constant supply of susceptible individuals. If the population is too small to maintain endemic transmission then the virus will be eliminated, and the critical size is 250,000–500,000 naïve individuals. In 1968, a very safe and effective measles vaccine became available and this led to a dramatic decline in the incidence of the disease and fatalities from it. Over 2.6 million people died from the disease in 1980 but by 2014 this had been reduced to 73,000. Unfortunately, due to the rise of misplaced anti-vaccination campaigns, the numbers of naïve individuals are coming close to the critical level in many industrialized countries (Figure 25.1). Unlike measles virus, canine distemper virus can be maintained in target species living in low-density populations and, as we shall see later, there are reservoir species.

Figure 25.1. Cases of measles in Europe January–August 2018.

Source: World Health Organization via Statista. Reproduced under Creative Commons License CC BY-ND 3.0 <https://www.statista.com/chart/15856/measles-cases-in-europe/>.

Before its global eradication through vaccination in 2011, rinderpest was a highly contagious disease affecting cattle, buffalo, yaks, and wild ungulates. The mortality rate was ~ 90% and complete herds of animals could be wiped out within a week of infection. Like measles virus, rinderpest needed large populations of susceptible animals for it to become endemic in particular areas and it is believed that its emergence as a problem coincided with the rise of farming somewhere in the east of Asia. The disease spread westwards as a result of invasions by the Huns and Mongols. Their armies were accompanied by large herds of cattle which provided draft power for their carts and a source of food. Curiously, the Asian Grey Steppe oxen used by the invaders were resistant to the effects of rinderpest but could spread the disease wherever they went. Baghdadlis was the name given to cattle traded to Egypt from Iraq in the early twentieth century and they were notorious for spreading rinderpest without themselves being seriously affected.

For many centuries no country in Europe was free from rinderpest. In the eighteenth century Pope Clement XI was so concerned about the effects of the disease upon the papal herds that he ordered his physician Dr Lancisi to do something to stop the spread. His recommendations were slaughter to reduce spread, restricted movement of cattle, burial of whole infected animals, and inspection of meat—recommendations that are appropriate today with other cattle infections such as foot and mouth disease. The massive losses of cattle in France led to the establishment of the world's first veterinary school at Lyons in 1762. Despite the barrier of the English Channel, Britain did not escape the disease as a result of the importation of infected cattle from the Netherlands in 1714.

Until the mid-nineteenth century, the major factor causing the spread of rinderpest was the movement of armies. However, the development of steam-powered ships and railways made possible the mass transport of cattle. This led to a Europe-wide pandemic between 1857 and 1867, largely because of the sale of infected cattle by the Russian Tsar who was running out of money. To make matters worse, in the nineteenth century the European powers were building their empires around the world and cattle were introduced to their colonies. In 1887, the Italians introduced infected cattle to Ethiopia via the port of Massawa (now in Eritrea). By 1891 the disease had crossed the African continent and reached the Senegal river and spread south into the western and eastern Rift valleys and thence all of east Africa. By 1896 it had got as far as South Africa. RNA sequence analysis of all known isolates of African rinderpest confirmed that they all evolved from the virus introduced to Ethiopia in 1887. Fortunately, in 1960, the English veterinary scientist Walter Plowright developed an inactivated vaccine that could be grown in tissue culture. This vaccine induced lifelong immunity without major side effects or the risk of further transmission and could be produced at a low cost. Its widespread use led to the global eradication of rinderpest.

The first clearly written description of a disease that resembles measles was made by Rhazes of Baghdad around 900 AD. He discriminated measles from smallpox and called it the 'smaller disease' hence the name given to the virus family: *morbilli* is the diminutive of *morbus*. The genome sequences of rinderpest and measles virus are very similar and it is believed that measles arose when rinderpest jumped the species barrier and infected humans. A number of researchers have tried to establish when this spill over occurred and the dates vary from the sixth century BC to the eleventh or twelfth century AD! The problem with trying to do molecular clock analyses with morbilliviruses is that there are no very old viral samples with which to calculate the rate of mutation. What we do know is that measles was absent from pre-Columbian populations in the Americas. However, once it was introduced by the early colonists it found populations with no immunity and mortality rates were ~ 70%.

Just as measles was absent from the Americas before the arrival of Columbus, canine distemper was absent from Europe. Dogs were introduced to the New World by Columbus and subsequently were widely used by colonists to detect natives who were waiting to ambush them. However, it was 1746 before distemper was first described in South America and the 1760s before returning dogs took the disease to Europe. There is a school of thought that distemper arose when the measles virus jumped into canines but this has never been proven. Unlike rinderpest and measles, canine distemper virus is promiscuous in its choice of hosts and these include ferrets, tigers, lions, and non-human primates among others. In 1994, approximately 30% of the lions in the Serengeti National Park in Tanzania were killed by canine distemper virus. This outbreak was traced to domestic dogs and since then there has been widespread vaccination of dogs in the vicinity of the park. Whilst the incidence of distemper in dogs has dropped, outbreaks in lions are still occurring suggesting that there are other reservoirs of the disease.

Canine distemper virus is not restricted to terrestrial mammals as it has been found to be widespread in aquatic species as well where it is known as phocine distemper. In 1987 it was found to have caused mass mortality in the freshwater seals found in Lake Baikal and in 2000 in seals in the Caspian Sea. These seals are thought to have been infected through contact with either domestic dogs or wolves. Introduction of canine distemper virus into Antarctic phocid seals is thought to have been via infected sled dogs imported from Arctic Canada and Greenland where the disease is endemic. The virus also is prevalent in sea lions in many other parts of the world.

Insights into the host specificity of morbilliviruses has been provided by genome sequencing. As noted in Box 25.1, the virus initiates an infection

Table 25.1 Morbillivirus specificity for two key cellular proteins.

Virus	Slam	Nectin 4
Rinderpest	bovine	bovine
Measles virus	human, non-human primates, canine	Human, non-human primates
Canine distemper virus	Canine and others	canine
Peste de petits ruminants virus	ovine	ovine
Phocine distemper virus	canine	human

when its H protein binds to the signalling lymphocyte activation molecule (SLAM) on the surface of lymphocytes and is released when infected cells bind to nectin 4 on epithelial cells. Table 25.1 shows the specificity of the different morbilliviruses for SLAM and nectin 4. What Table 25.1 does not show is how little sequence change is necessary to alter these specificities. For example, a single amino acid change in the H protein of canine distemper virus or peste de peitits ruminants virus allows them to bind to the human SLAM. This illustrates how easy it could be for morbilliviruses to jump species and there have been incidences when peste de petits ruminants virus has jumped from goats to cattle and camels. These findings also give credence to the thoughts that measles virus resulted from species jumping by rinderpest and canine distemper by spill over from measles-infected humans. They also show how easy it is for zoonoses to evolve.

Key points

- Morbilliviruses are RNA viruses that infect humans and a wide variety of domestic and wild animals. They are highly infectious, spread via the respiratory route, cause immune suppression, and are associated with high morbidity and mortality. The best-known examples are measles, rinderpest, and canine distemper.

- Rinderpest was a disease of cattle and other ungulates. It is believed to have evolved in Asia around the time that humans transitioned from hunter gatherers to farmers. Invading armies brought the disease to the West with their animals. The European colonialists then took the disease to Africa. The disease now has been eliminated by vaccination.

- Measles is highly infectious and one of the leading causes of death among young children. Maintenance of measles virus in a population requires a constant supply of susceptible individuals. Vaccination is very effective at preventing the disease but misplaced anti-vaccination campaigns are resulting in a surge in cases.

- Measles was introduced to the Americas, with devastating effects, by early colonialists. By contrast, canine distemper was absent from Europe in the pre-Columbian era. The dogs of colonialists caught the virus in the Americas and brought it to Europe.

- It is believed that measles arose when rinderpest jumped species. Sequence analysis of different morbilliviruses shows that it would take only a few mutations for them to jump species.

Suggested Reading

Abdullah N., Kelly J.T., Graham S.C., Birch J., Goncalves-Carneiro D., et al. (2018) Structure-guided identification of a nonhuman morbillivirus with zoonotic potential. *Journal of Virology* **92**(3), e01248-18. doi:10.1128/JVI.01248-18

Dux A., Lequime S., Patrono L.V., Vrancken B., Boral S., et al. (2020) Measles virus and rinderpest divergence dated to the rise of large cities. *Science* **368**, 1367–70.

Fukuhara H., Ito Y., Sako M., Kajikawa M., Yoshida K., et al. (2019) Specificity of morbillivirus haemagglutinins to recognize SLAM of different species. *Viruses* **11**(8), 761. doi: 10.3390/v11080761

Kennedy J.M., Philip Earle J.A., Omar S., Abdullah H., Nielsen O., et al. (2019) Canine and phocine distemper viruses: global spread and genetic basis of jumping species barriers. Viruses 11(10), 944. doi:10.3390/v11100944

Nambulli S., Sharp C.R., Acciardo A.S., Drexler J.F., and Duprex W.P. (2016) Mapping the evolutionary trajectories of morbilliviruses: what, where and whither. *Current Opinion in Virology* **16**, 95–105.

Viana M., Cleaveland S., Matthiopoulos J., Halliday J., Paker C., et al. (2015) Dynamics of a morbillivirus at the domestic-wildlife interface: canine distemper virus in domestic dogs and lions. *Proceedings of the National Academy of Sciences* 112, 1464–9. doi: 10.1073/pnas.1411623112

26

Filovirus Haemorrhagic Fevers: Marburg Virus and Ebola

In August 1967, a new disease caused alarm in the world of virology when a small outbreak occurred in laboratory workers in Germany. The disease began in the first patients with symptoms of extreme malaise, myalgia, headache, and a rapid increase in body temperature. Initially, the symptoms were not very alarming but soon the condition of the patients worsened and they had to be hospitalized. Many of the patients developed haemorrhagic fever (Box 26.1) and seven patients died. A total of thirty-two people were infected, thirty in Germany (twenty-four in Marburg and six in Frankfurt), and two in Belgrade (Serbia). Of these, twenty-six were primary infections and the common factor was that all had had contact with the blood, organs, or cell-cultures from a batch of African green monkeys imported from Uganda. The six secondary infections were caused by such things as needle-stick injuries while handling patient blood samples. Subsequently it was shown that the patients were infected with a hitherto unknown virus that was given the name Marburg virus. This was the first isolate of a new group of viruses that were given the name filoviruses on account of their filamentous morphology.

Box 26.1 VIRAL HAEMORRHAGIC FEVERS

Viral haemorrhagic fevers are a diverse group of animal and human illnesses in which fever and haemorrhage are the major symptoms. These diseases most commonly occur in tropical areas of the world. When viral haemorrhagic fevers occur in temperate regions, they are usually found in people who recently have travelled internationally. Several distinct families of viruses cause haemorrhagic fevers including filoviruses (Marburg, Ebola), flaviviruses (yellow fever, dengue), and arenavirus (Lassa fever).

Microbiology of Infectious Disease. Sandy B. Primrose, Oxford University Press.
© Sandy B. Primrose (2022). DOI: 10.1093/oso/9780192863843.003.0026

Viral haemorrhagic fevers (VHFs) typically manifest as rapidly progressing acute febrile syndromes with profound haemorrhagic complications. The latter is the result of the virus infection damaging the walls of blood capillaries, making them leaky. Infection can also interfere with the blood's ability to clot. The internal bleeding that results can range from relatively minor to life-threatening.

Outbreaks of viral haemorrhagic fevers are associated with an enormous impact in terms of human lives and costs for the management of cases, contact tracing, and containment. Unfortunately, surveillance, diagnostic capacity, infection control, and the overall preparedness level for management of such fevers are very limited in most endemic countries. The availability of appropriate protective equipment and education of healthcare workers about safe clinical practices and infection control is the mainstay for the prevention of the spread of such viral diseases.

Since the initial outbreak of Marburg virus in 1967, outbreaks and sporadic cases have been reported in Angola, Democratic Republic of Congo, Kenya, and South Africa (in a person who had recently travelled to Zimbabwe). The largest outbreak on record to date occurred in 2005 in Angola and involved 374 patients of whom 329 died. The most recent cases were from a small outbreak in Uganda in 2017. Many of the outbreaks started with men who were working in bat-infested mines (see Box 26.2). The virus then got transmitted within their communities through sexual intercourse or exposure to clothing and bedding contaminated with body fluids from infected individuals. Two unrelated sporadic cases in travellers occurred during 2008 following visits to a tourist attraction: the 'python cave' in the Maramagambo Forest in western Uganda. This cave also is home to a large colony of Egyptian fruit bats. Both people became ill after returning to their home country, one in the Netherlands and one in the United States.

Egyptian fruit bats (*Rousettus aegyptiacus*) have a wide distribution in Africa and occur in all the locations where there have been outbreaks of Marburg virus (Figure 26.1). In 2009, Marburg virus was isolated from healthy specimens of these bats suggesting that they play a role in outbreaks. However, it is not clear whether these bats are the actual reservoirs of the disease of if they are intermediate hosts that get infected by contact with another animal reservoir. If bats free of the Marburg virus are infected experimentally then they remain healthy even though the virus replicates to high titres in organs such as the liver and spleen. The African green monkeys that had caused the original outbreak are not reservoirs of the disease but intermediate hosts.

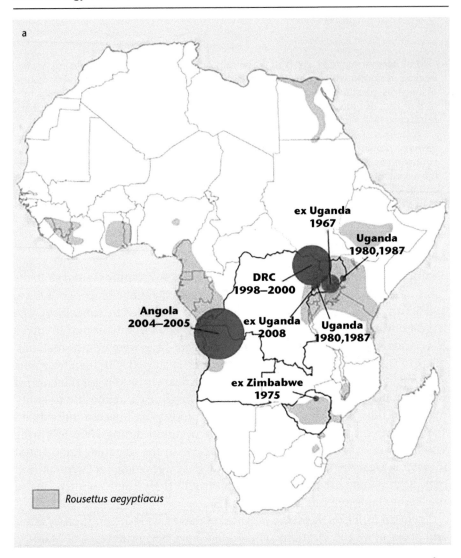

Figure 26.1. Location of Marburg virus (MARV) infections (circle sizes correspond to reported number of MARV cases) and distribution of the Egyptian fruit bat (*Rousettus aegyptiacus*) in Africa.

Source: Kristina Brauburger, Adam J. Hume, Elke Mühlberger, and Judith Olejnik/Wikimedia Commons, reproduced under Creative Commons Attribution 3.0 Unported license <https://commons.wikimedia.org/wiki/File:Viruses-04-01878-g001-A.jpg>.

Box 26.2 BATS AS RESERVOIRS OF VIRUSES

With 1,400 species, bats are the second largest group of mammals (after rodents) and the only ones capable of true and sustained flight. Most bats are nocturnal and many roost in caves or other refuges that are dark during daytime. Bats are natural reservoirs for a large number of zoonotic pathogens including many viruses and they seem to be resistant to the pathogens that they carry. Their high mobility, broad distribution, long life spans, and social behaviour make bats favourable hosts and vectors of disease.

Many bats are insectivores, and most of the rest are frugivores (fruit-eaters) or nectarivores (nectar-eaters). A few species feed on animals other than insects and some, such as the vampire bats, feed on blood. These feeding habits can result in transmission of viruses to humans. For example, in the 1930s it was shown that rabies is endemic in vampire bats in Trinidad and can transmit the virus to humans whilst feeding on blood. Eating unwashed fruit has resulted in the transmission of viruses from flying foxes (fruit bats) to humans. Nipah virus transmission has been linked to the consumption of date palm sap after bats fed on it. Bat dung is rich in nitrates and is mined from caves for use as a fertilizer. Unfortunately, bats can excrete viruses in this dung resulting in infections of miners as has happened with Marburg virus infections (see main text).

Sequence analysis of the genomes of Marburg virus isolates from the different outbreaks and determination of the number of nucleotide substitutions suggest that the various isolates had a common ancestor about 180 years ago in the mid-nineteenth century, most probably in Uganda. However, there could have been outbreaks earlier than that but, if there were, they would have been confined to particular localities because of the absence of infrastructure at the time.

Fortunately, recorded cases of Marburg virus disease are rare and the existence of the disease is largely unknown to the general public. Not so disease caused by the next filovirus to be discovered: Ebola virus. Unlike Marburg virus, Ebola virus has a case fatality rate of 60% to 80%. In 1976 there were two almost simultaneous outbreaks of a fatal haemorrhagic fever in central Africa, one near the Ebola River in Zaire (now Democratic Republic of Congo) and the other 500 miles away in what now is South Sudan (Figure 26.2). Initially it was thought that the two outbreaks were caused by an infected person who travelled between the two locations. Subsequently it was shown that the two outbreaks were caused by genetically distinct viruses (*Zaire ebolavirus* and *Sudan ebolavirus*) and so had different origins.

The two initial outbreaks of Ebola virus were followed by a small outbreak in 1979 after which the virus seemed to disappear. Imagine then the panic in 1989 caused by the discovery of an Ebola virus (*Reston ebolavirus*) in monkeys in a research facility at Reston, Virginia. Workers in the facility were all too aware of the initial outbreak of Marburg virus in scientists handling African green monkeys and of the high fatality in the African cases of Ebola virus disease. Of particular concern was the fact that this virus spread through the monkey colony by aerosols. The monkeys had come from the Philippines and subsequently the virus was found in sick pigs in that country. Workers handling the monkeys were found to have developed antibodies to the virus but fortunately none had any disease symptoms.

Ebola virus disease re-emerged in 1994 with the first of fifteen new outbreaks caused by *Zaire ebolavirus* (Figure 26.2). There also have been seven further outbreaks of disease due to *Sudan ebolavirus*. A new species of Ebola virus, *Tai Forest ebolavirus*, was discovered in west Africa in 1994 in a veterinarian who had been conducting an autopsy on a chimpanzee. This patient recovered and is the only person known to be infected with *Tai Forest ebolavirus*. In 2007, yet another new Ebola virus, *Bundibugyo ebolavirus*, was discovered in Uganda. This is the least lethal species of Ebola virus with a case fatality rate of 38%. A sixth species of Ebola virus (*Bombali ebolavirus*) was discovered in bats in Sierra Leone in 2018 but it is not known if this can cause disease in humans.

Until 2013, all the outbreaks of Ebola virus disease occurred in rural areas and mostly in central Africa (Sudan, Uganda, Gabon, Republic of Congo, and the Democratic Republic of Congo). The number of cases in each outbreak was less than 600 and so the disease attracted little attention except from public health officials and a few virologists. Then, in 2014, a major outbreak of *Zaire ebolavirus* occurred in west Africa for the first time. This outbreak is notable in that it was larger in scale than all the other outbreaks combined with nearly 30,000 cases and over 11,000 deaths. It started in south-east Guinea but soon spread to neighbouring Sierra Leone and Liberia. Seven other countries had a small number of cases but all were linked to west Africa. Subsequent analysis showed that viral load was a major factor in determining if patients died from the disease. The virus has an RNA genome and is replicated by a virus-encoded RNA-dependent RNA polymerase. This is error-prone and many survivors were infected with virus variants with mutations in the gene encoding the polymerase.

A number of factors contributed to the high incidence of disease in the 2014–2016 outbreak. First, the local population are highly mobile and move freely, not just between the three countries but between rural areas and cities. Second, traditional burial rituals where mourners touch the bodies of the dead resulted in a high percentage of onward infections. Another significant funeral practice, in accordance with Muslim tradition, was for female relatives to wash the body of women who had died of Ebola. Finally, a lack of adequate healthcare facilities combined with powerful traditional healers meant that many patients were not treated in a way that would stop spread of the disease. Were it not for the outstanding actions of the charity Médecins Sans Frontières, later supported by teams from the World Health

Organization, the outbreak would not have ended in 2016. Following an Ebola outbreak in 2018 in the Democratic Republic of the Congo, a socio-logical study revealed that many people did not accept what they were told about transmission of the disease. Rather, they believed disease was spread by politicians or multinational corporations. As long as these beliefs persist there will be outbreaks.

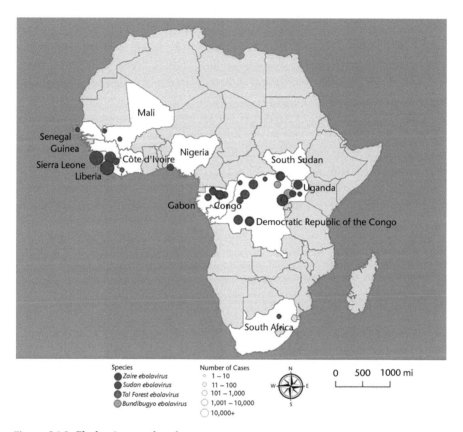

Figure 26.2. Ebola virus outbreaks.

Following the 1976 outbreaks of Ebola, scientists began looking for the natural reservoir of the virus. They screened thousands of animals including gorillas, chimpanzees, elephants, fruit bats, and insects but all to no avail. In fact, Ebola is as much a threat to chimpanzees and gorillas as it is to humans so they are not the reservoir. Retrospective analysis of the 2014–2016 out-break traced the source to a village called Meliandou in south-east Guinea. Primates are rare in this area and, as the first cases of Ebola were in women and children, game animals are not likely to be the reservoir. The index case in Meliandou was a two-year-old boy and he and others used to play in a hollow tree close to the village. By the time epidemiologists arrived in Meliandou the tree had burned down but villagers reported that it had been full of insectivorous bats. These insectivorous bats, as opposed to fruit bats, can harbour Ebola virus based on PCR and serological analysis and may have been the source of the outbreak. However, unlike Marburg virus, live Ebola virus has not been isolated from bats.

Sequence based analysis of Ebola virus genomes from different patients in the 2014–2016 outbreak suggests that they shared a common ancestor around the beginning of 2014. This fits with the reports that the index case in Meliandou fell ill and died in December 2013. Analysis of isolates from all outbreaks of *Zaire ebolavirus* had a common ancestor as recently as 1975, which is close to the first recorded outbreak, even though Zaire and west Africa are thousands of miles apart. When a comparison is made of the Zaire, Sudan, Tai Forest, Reston, and Bundibugyo ebolaviruses, they appear to have shared a common ancestor about 2,000 years ago. Interestingly, there has been a suggestion that Ebola virus was responsible for an outbreak of disease in Athens that was recorded by the historian Thucydides in 430 BC. Remnants of filoviruses have been found in the genomes of a variety of mammals as diverse as rats and marsupials indicating that filoviruses have been around for a very long time.

Key points

- Filoviruses are a family of viruses that infect mammals in tropical regions and cause haemorrhagic fevers when they infect humans. The best-known members of the family are Marburg virus and Ebola.

- The first outbreak of Marburg virus occurred in laboratory workers handling African green monkeys from Uganda. Subsequent outbreaks occurred in Africa among workers mining guano from bat-infested caves. It is not known if monkeys and bats are reservoirs of the virus or intermediate hosts even though they show no symptoms when infected.

- There have been over twenty outbreaks of Ebola in various parts of Africa in the last twenty-five years. The virus is highly infectious and spread is facilitated by close contact, especially local burial practices. A refusal to accept the role of virus in the disease makes control very difficult.

- The different regional outbreaks of Ebola have been caused by different strains of the virus and sequence analysis suggests they had a common ancestor 2,000 years ago. However, all attempts to find the animal reservoir have failed.

Suggested Reading

Coltart C.E.M., Lindsey B., Ghinai I., Johnson A.M., and Heymann D.L. (2017) The Ebola outbreak, 2013–2016: old lessons from new epidemics. *Philosophical Transactions of the Royal Society of London, B: Biological Sciences* **372**, 20160297. doi:org/10.1098/rstb.2016.0297

De La Vega, M.-A., Stein D., and Kobinger G.P. (2015) Ebolavirus evolution: past and present. *PLoS Pathogens* **11**(11), e1005221. doi:10.1371/journal.ppat.1005221

Di Paola N., Sanchez-Lockhart M., Zeng X., Kuhn J.H., and Palacios G. (2020) Viral genomics in Ebola virus research. *Nature Reviews Microbiology* **18**, 365–78. doi:org/10.1038/s41579-020-0354-7

Holmes E.C., Dudas G., Rambaut A., and Andersen K.G. (2016) The evolution of Ebola virus; insights from the 2013–2016 epidemic. *Nature* **538**, 193–200

Kazanjian P. (2015) Ebola in antiquity? *Clinical Infectious Diseases* **61**, 963–8

Letko M., Seifert S.N., Olival K.J., Plowright R.K., and Munster V.J. (2020) Bat-borne virus diversity, spillover and emergence. *Nature Reviews Microbiology 18*, 461–71. doi:org/10.1038/s41579-020-0394-z

Muzembo B.A., Ntontolo N.P., Ngatu N.R., Khatiwada J., Ngombe K.L., et al. (2020) Local perspectives on Ebola during its tenth outbreak in DR Congo; a nationwide qualitative study. *PLoS One* **15**(10), e0241120. doi:10.1371/journal.pone.0241120

Olejnik J., Muhlberger E., and Hume A.J. (2019) Recent advances in Marburgvirus research. *F1000 Research* **8**, 704. doi:10.12688/f1000research.17573.1.ecollection 2019

Saéz A.M., Weiss S., Nowak K., Lapeyre V., Zimmerman F., et al. (2015). Investigating the zoonotic origin of the West Africa Ebola epidemic. *EMBO Molecular Medicine* **7**, 17–23.

27

The Origins of HIV and the AIDS Pandemic

There is probably no adult in the Western world who has not heard of acquired immunodeficiency syndrome, more simply known as AIDS. Thus, it may come as a surprise that this disease was unknown until forty years ago when it appeared in the gay community in the United States. At that time there was no cure for the disease and since diagnosis was a death sentence, articles about the origins of this mystery disease and its link to sexual practices were regular features in newspapers and scientific and medical journals.

The first indications of a hitherto unknown disease emerged when there were clusters of two rare conditions in homosexual men. Pneumocystis pneumonia (PCP) is a form of pneumonia caused by the fungus *Pneumocystis jirovecii*. This organism is an opportunistic pathogen that only infects people with impaired immunity but in 1981 it appeared in a group of injecting drug users and gay men with no known cause of immunodeficiency. Around the same time there were reports from California and New York of clusters of Kaposi's sarcoma (KS) in homosexual men. KS is a type of cancer that occurs in people with a weakened immune system. Soon there was a significant increase in the numbers of cases of PCP and KS and this led to the involvement of the US Centers for Disease Control and Prevention (CDC). Initially they called it the '4H disease' because it seemed to affect heroin users, homosexuals, haemophiliacs, and recent immigrants from Haiti. Soon it became apparent that the disease affected both men and women and that it could be transmitted through unprotected sex, blood transfusions, and use of contaminated syringe needles. Consequently, in 1982, CDC started referring to the disease as AIDS.

The rapid rise in the number of cases of AIDS strongly suggested that the causative agent was a microorganism and so it proved to be. In 1983, two different research groups led by Luc Montagnier in France and Robert Gallo in the United States reported that they had isolated retroviruses (see Box 27.1) from patients with AIDS. Subsequently it was shown that the two

Microbiology of Infectious Disease. Sandy B. Primrose, Oxford University Press.
© Sandy B. Primrose (2022). DOI: 10.1093/oso/9780192863843.003.0027

Box 27.1 RETROVIRUSES

A retrovirus is a virus that has an RNA genome and carries a number of enzymes including reverse transcriptase and integrase. After entering a cell, the reverse transcriptase converts the viral RNA into DNA and this DNA is incorporated into the host cell genome by the integrase. The integrated retroviral DNA is referred to as a provirus. The host cell treats the viral DNA as part of its own genome, transcribing and translating the viral genes along with the cell's own genes, producing the proteins required to assemble new copies of the virus.

There are three basic groups of retroviruses. Lentiviruses (slow retroviruses) such as HIV and SIV are characterized by long incubation periods in mammals and cause severe immunodeficiency and death in humans and other animals. The oncogenic retroviruses are able to cause cancer in some species and the spumaviruses (foamy viruses) are benign and not linked to any disease in humans or animals.

Six different classes of anti-retroviral drug have been developed to treat patients with HIV/AIDS and these usually are administered as combinations of two or three different drugs. The reason for this is that the enzyme reverse transcriptase shows a low fidelity of copying RNA into DNA and this results in the generation of many mutants, some of which are resistant to the drugs.

---

virus isolates were the same virus and in 1986 it was given the name human immunodeficiency virus (HIV). Once the viral cause of AIDS had been identified it was possible to investigate the mechanism of disease and it transpired that HIV infects vital cells in the human immune system, such as helper T cells (specifically $CD4^+$ T cells), macrophages, and dendritic cells. When $CD4^+$ T cell numbers decline below a critical level, cell-mediated immunity is lost, and the body becomes progressively more susceptible to opportunistic infections such as PCP, human gammaherpesvirus 8 (HHV-8) which causes KS, and other diseases such as tuberculosis. Today, two types of HIV are recognized: HIV-1 and HIV-2. HIV-1 is the virus initially isolated from patients with AIDS. It is more virulent and more infective than HIV-2 and is the cause of the majority of HIV infections globally. HIV-2 has a relatively poor capacity for transmission and is largely confined to central Africa.

It is estimated that 38 million people across the globe have HIV/AIDS today and that roughly the same number of people have died from it since its identification in the early 1980s. This begs the question as to what caused its sudden emergence and epidemic spread.

An early clue came in 1986 with the isolation of HIV-2 from AIDS patients in west Africa. HIV-2 was shown to be distantly related to the original HIV-1 isolate but was closely related to a simian virus that caused immunodeficiency in captive macaques. Soon related viruses, termed simian immunodeficiency viruses (SIVs), were found in various species of African monkeys and apes (Table 25.1) although the animals often showed no signs of disease. Macaques are native to Asia rather than Africa and SIV is not

Table 27.1. Examples of animals that may be naturally infected with SIVs. The incidence of infection varies from < 1% to > 50%.

Non-Human Primate	Designation
Chimpanzee	SIVcpz
Western gorilla	SIVgor
Red-capped mangabey	SIVrcm
Mandril	SIVmnd
L'Hoest's monkey	SIVlho
Vervet monkey	SIVver
Mantled guereza	SIVcol
Syke's monkey	SIVsyk
Sooty mangabey	SIVsmm
Western red colobus monkey	SIVwrc

found in wild specimens but the captive macaques that developed immunodeficiency had been inoculated with blood from a sooty mangabey (an African primate). Subsequently, a simian relative of HIV-1 was found in chimpanzees. This suggested that HIV strains arose when SIVs infected humans (see Table 27.1).

There are many barriers to viruses crossing between species but genetic relatedness increases the likelihood. Chimpanzees and humans have approximately 98.8% DNA sequence identity. This, and the close genetic relationship between the SIV infecting chimpanzees (SIVcpz) and HIV-1, suggests that the original HIV-1 strain arose from a successful infection of a human with SIVcpz. This would have required blood from an infected chimpanzee to have come in contact with mucous membranes or a cut in the skin of a human. Such an event could easily have occurred during hunting for bushmeat which is a common practice in central and west Africa. By looking at the divergence in the RNA sequences of SIVcpz and HIV-1 it is estimated that the jump across species occurred about 100 years ago in central Africa, probably in south-eastern Cameroon (see Box 27.2).

Once the SIV had made the leap into a human it would have been spread in the local community by sexual contact. In this context it is worth noting that in sub-Saharan Africa, HIV/AIDS is spread predominantly by heterosexual practice whereas in the northern hemisphere it is spread mainly by homosexual practices. Around the time of the initial transmission event there was a rapid growth of large colonial cities such as Brazzaville and Kinshasa (formerly Leopoldville) that sit on opposite sides of the Congo River. The growth of these cities resulted in mass migration of people from the bush, an event accompanied by an increase in sexual promiscuity and prostitution. Rivers, which serve as major routes of travel in Africa, would have provided the means whereby HIV spread from south-eastern Cameroon to the major cities of the Republic of Congo and the Democratic Republic of Congo (formerly Zaire) (Figure 27.1). We know that the virus was present in Kinshasa (Democratic Republic of Congo) in the late 1950s and early 1960s because it has been detected in blood and serum samples collected at that time. Also, a near full-length genome of HIV-1 was isolated recently from a tissue specimen that was formalin-fixed and paraffin-embedded in that year.

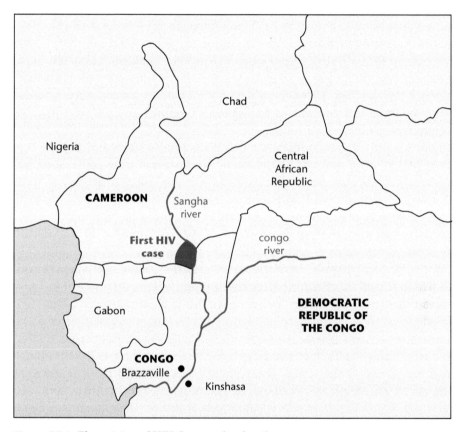

Figure 27.1. The origins of HIV. See text for details.

Box 27.2 MOLECULAR CLOCK ANALYSIS

Many viruses, especially RNA viruses, rapidly accumulate genetic variation because of short generation times and high mutation rates. This is especially true with lentiviruses such as SIV and HIV because the essential enzyme reverse transcriptase is particularly

continued

Box 27.2 *Continued*

---

error-prone. Changes in RNA sequences accumulate continuously in the genome over time, so comparing the RNA sequences between SIVcpz and HIV-1 allows us to esti- mate the time since they last shared a common ancestor. This analysis suggests that the most recent common ancestor of the two viruses existed between 1910 and 1930.

Since the original HIV-1 type arose it has undergone genetic change and nine subtypes (A–D, F–H, J, K) have been identified. Subtypes A and D origi- nated in central Africa and eventually spread to eastern Africa. Subtype C was introduced to South Africa from where it spread to India and Asia. Subtype B predominates in Europe and the Americas but its route there is of particular interest. In 1960 the Democratic Republic of Congo got its independence from Belgium and this led to an efflux of administrators. Unfortunately, Belgium had not trained local people to replace them and so the United Nations recruited Francophone experts from around the world to help the fledgling country. Around ,500 of these experts came from Haiti. Some of these Haitians acquired HIV subtype B during their stay in Africa and at least one took it back to their home country in the late 1960s.

Immediately after becoming infected with HIV, a patient may display influenza-like symptoms or develop a body rash and any Haitian who devel- oped these symptoms in Africa would have assumed that they were suffering from some local disease. After the initial infection, there is a latent period of five to fifteen years before AIDS develops and so the disease would have spread silently in the Haitian population. This explains why there was high incidence among Haitian immigrants in the United States soon after AIDS was recognized as a disease. However, these Haitian immigrants were not re- sponsible for the epidemic spread of AIDS even though thousands of Haitians had the disease by the time it was recognized in the early 1980s. Rather, in the early 1970s, Haiti was a popular destination for 'sexual tourism' and this led to the introduction of subtype B to New York city. The virus then spread rapidly among sexually promiscuous homosexual men and by 1976 it was circulating in the gay community in San Francisco as well. Soon thereafter it spread throughout North America and Europe (Figure 27.2). For a long time, it was believed that the person who introduced HIV into North America, the so-called patient zero, was a promiscuous French-Canadian flight attendant but this now has been disproved.

Today, four groups of HIV-1 are recognized and each arose from an inde- pendent transmission event. Groups M and N arose from SIVcpz and groups O and P from SIVgor. Interestingly, SIVgor arose when SIVcpz crossed from chimpanzees into gorillas. Similarly, there were nine independent transmis- sion events from sooty mangabeys (SIVsmm) to humans giving rise to HIV-2

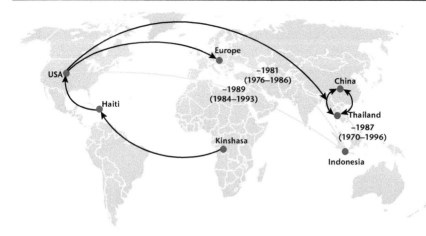

Figure 27.2. The spread of HIV from Africa.
Source: <https://www.eurekalert.org/multimedia/pub/213181.php>

groups A to I. However, HIV-1 group M (for 'major') is the only one that is pandemic and accounts for > 98% of cases of HIV/ AIDS. So, why have the nine HIV-2 groups and HIV-1 groups N, O, and P not been successful human pathogens?

When viruses jump from one species to another, they encounter a whole panoply of antiviral defence systems that they need to overcome if they are to become established in their new host. These antiviral systems are very effective and so cross-species transmission of viruses is a rare event. It is likely that humans hunting bushmeat regularly got infected with SIV but very few of these infections resulted in human disease. On the rare occasions when an SIV did establish itself in humans it would have remained as a very local disease until the development of cities with their burgeoning populations enabled a more general spread of infection. However, a key event was the development of certain mutations in a viral protein called Vpu. This enabled the virus to overcome a key host defence protein called tetherin which blocks many viruses from leaving infected cells. HIV-1 group M, and only group M, has acquired these mutations and given it the capacity to spread pandemically.

Key points

- There are two types of human immunodeficiency virus: HIV-1 and HIV-2. There are four sub-types of HIV-1 and sub-type M is responsible for the global AIDS pandemic. There are nine sub-types of HIV-2.

- HIV-1 arose when an immunodeficiency virus jumped species from a chimpanzee to a human in Cameroon at the beginning of the twentieth century. The virus circulated locally through sexual intercourse.

- Movement of workers took the virus to Kinshasa in the Belgian Congo where it was spread through sexual intercourse. An influx of Haitian administrators following independence of the Belgian Congo in 1960 led to the virus being taken to Haiti. As a result of sexual tourism, HIV was taken to the United States in the 1970s. This led to the current AIDS pandemic.

- The four sub-types of HIV-1 were the result of spillover from chimpanzees (twice) and gorillas (twice). The nine sub-types of HIV-2 resulted from spillover from sooty mangabeys. HIV-1 sub-type M has a particular mutation not found in all the other HIV sub-types and which allows it to overcome host defences. This explains its predominance in individuals with AIDS.

Suggested Reading

Sauter D. and Kirchhoff F. (2019) Key viral adaptations preceding the AIDS pandemic. *Cell Host & Microbe* **25**, 27–38. doi:org/10.1016/j.chom.2018.12.002

Sharp P.M. and Hahn B.H. (2011) Origins of HIV and the AIDS pandemic. *Cold Spring Harbor Perspectives in Medicine* 1:a006841

Worobey M., Watts T.D., McKay R.A., Suchard M.A. et al. (2016) 1970s and 'Patient 0' HIV-1 genomes illuminate early HIV/AIDS history in North America. *Nature* **539**, 98–101. doi:10.1038/nature19827

28

The Benefits of a Segmented Genome: Influenza

In 1918, just as the end of the First World War was in sight, a deadly respiratory disease spread around the world in three consecutive waves. Approximately one-third of the world's population got infected and it is estimated that 50 million people died from the disease—which puts the recent COVID-19 pandemic in perspective. Unlike COVID-19 which affected mostly older people, the majority of victims in the 1918 pandemic were young adults who previously had been healthy. During the 1918 pandemic, the respiratory route of transmission was clearly identified. Also, the disease was shown to be caused by an agent that could pass through a filter that would remove bacteria and so probably was a virus, the one that we now know as influenza. In 1931, Richard Shope identified a virus as the cause of swine influenza, a disease of pigs that was identified at the same time as the human pandemic. His discovery led to the isolation of human influenza viruses two years later. It also provided the key to the origin of influenza pandemics.

Three influenza pandemics have occurred since the Second World War, each less severe than the 1918 pandemic. The first of these was the Asian influenza pandemic that started in Yunnan province in China in 1957 and probably accounted for 1 million deaths. The next pandemic occurred in 1968 and began in Hong Kong before causing 500,000 to 2,000,000 deaths. This was the first pandemic to spread significantly by air travel. The most recent influenza pandemic was the 2009 'swine flu' outbreak which originated in Mexico and resulted in over 1 billion people being infected and nearly 300,000 deaths. Much less virulent strains of the influenza virus were in circulation in the years between these pandemics, and still are circulating, causing what is known as 'seasonal flu'. In a typical non-pandemic year, 5% to 15% of the population will contract influenza and there will be mortality, particularly among the elderly and people with chronic health

Microbiology of Infectious Disease. Sandy B. Primrose, Oxford University Press.
© Sandy B. Primrose (2022). DOI: 10.1093/oso/9780192863843.003.0028

conditions. So, what happens to turn seasonal influenza into pandemic influenza?

There are four main types of influenza, A, B, C, and D, and they are differentiated on the basis of the antigenicity of their nucleoprotein. Type A viruses infect a wide variety of birds and mammals and, as we shall see, this is a key factor in the origins of pandemics. Type B viruses also cause seasonal flu in humans while type C viruses are relatively benign. Type D viruses infect cattle but not humans. Influenza virus particles are surrounded by an envelope which is derived from the host cell membrane during the process of virus release from infected cells. Embedded in this envelope are three proteins and the two most important in the epidemiology of influenza are haemagglutinin (HA) and neuraminidase (NA). Based on their antigenicity, sixteen different HA sub-types (H1–H16) and nine different NA sub-types (N1–N9) have been found in avian and human influenza strains.

The HA plays a key role in the ability of influenza A to infect new hosts because it is responsible for binding of the virus to receptors on potential host cells. Avian influenza viruses preferentially bind to receptors in which sialic acid is joined to galactose via an α-2,3-linkage whereas for human influenza viruses, the receptors have a α-2,6-linkage. The latter are predominant in the upper respiratory tract of humans, including the epithelial cells lining the nasal cavity. Pigs and quails are unusual in having both receptors in their upper respiratory tract and this enables them to be simultaneously co-infected with avian and human viruses. The number of other mammals and birds that have both receptors is not known.

The function of the neuraminidase (NA) is to remove sialic acid residues from viral glycoproteins and infected cells during both entry and release of the virus from cells. Influenza virus particles with low NA enzymatic activity cannot be efficiently released from infected cells, resulting in the accumulation of large aggregates of progeny virions on the cell surface. Since the formation of aggregates results directly from HA binding to sialic acid receptors on cellular and viral surfaces, a balance of HA and NA activities is critical for successful infections. In other words, there should be enough HA activity to ensure virus binding and enough NA activity to ensure the release of progeny virus.

Influenza viruses have a single-stranded RNA genome that is segmented. In the case of type A and type B viruses there are eight RNA segments and the HA and NA genes are on different segments. As noted in earlier chapters, RNA-dependent RNA polymerase does not have a proofreading function and during viral RNA replication mutations will occur. These mutations result in a gradual change in different viral proteins with a concomitant change in disease severity and the susceptibility of particular individuals. This process is known as antigenic drift and it is responsible for the changes in the strains of seasonal influenza that circulate from one year to the next.

Occasionally, two different influenza viruses will infect the same cell. In this situation, the eight RNA segments from each of the two viruses (X1–X8 and Y1–Y8) will reassort independently and the progeny will be a mixture of 2^8 (i.e. 256) different combinations of X and Y. Not every possible combination of X and Y may occur or survive. However, among those that do survive there could be ones with novel epidemic potential, especially if the two input viruses are significantly different, for example, avian and human influenza viruses. This evolution of novel influenza strains by genetic reassortment is known as antigenic shift. Reassortment occurs within A strains and within B strains but not between A and B strains.

Pandemic influenza results when a viral strain emerges with an HA protein to which few people have prior immunity and the source of these HA genes that are new to humans is the extensive pool of influenza viruses infecting wild birds. It is uncommon for humans to be infected directly with avian strains of influenza viruses or for human strains to infect birds. More often than not, reassortment of the segments of the avian and human influenza genomes occurs in an intermediate host which is susceptible to both strains of the virus. Pigs are such a host and the way that they are farmed means that they are exposed regularly to both humans and birds. Antigenic shift could occur in other animals whose potential to host both avian and human influenza viruses is not yet recognized. The live bird markets in China where a wide range of different species of birds (chickens, ducks, geese, pigeons, etc.) in close proximity and large numbers are traded daily could well provide the ideal breeding ground for new influenza viruses.

For a long time, attempts to unravel the origins of the 1918 pandemic were thwarted because there had been no retention of any virus stocks from the period., either before or during the pandemic. Once techniques for gene manipulation and genome sequencing became available it became possible to recover viral material from formalin-fixed and paraffin-embedded lung tissue from 1918 influenza victims and from a victim buried in Alaska's permafrost. The fully reconstructed virus was found to be of the H1N1 subtype. Although there is no viral material from the period before the 1918 pandemic, we do know something about the subtype of virus that was circulating. In the period 1889 to 1892, there was an influenza epidemic in Russia and analysis of stored patient serum samples from the period indicates that it was of type H3Nx. It remains undefined whether the 1918 H1N1 pandemic virus originated from this virus by reassortment, or if it was introduced by a direct zoonotic transmission event of an avian, swine or other influenza virus. Regardless, the H1N1 virus continued to circulate, causing seasonal epidemics, until 1957 when it reassorted with an avian H2N2 virus and caused the Asian flu pandemic (Figure 28.1) This virus (H2N2) circulated until 1968, when it reassorted again with the avian H3Nx virus to give Hong Kong flu (H3N2), which has caused seasonal epidemics ever since. In 1977 the H1N1 virus was reintroduced in the human population and co-circulated with H3N2 viruses until the influenza pandemic of 2009. At this time, it was replaced by another H1N1 virus which was the result of multiple reassortment events between avian, swine, and human influenza viruses.

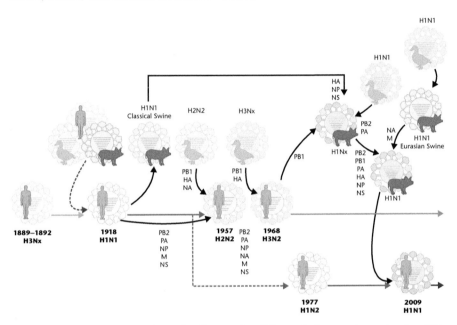

Figure 28.1. Reassortment events leading to the different influenza A pandemics. PB1, PB2, and PA are three different components of the viral RNA polymerase which are encoded on different RNA segments. NP is the nucleoprotein gene, M is the gene for a matrix protein, NS is a gene for a non-structural protein. Figure reproduced from Short, Kedzierska, and van de Sandt (2018).

The generation of human influenza viruses with a new HA component by antigenic shift is not sufficient to explain the pandemic potential and the lethality of influenza. In 2003 there was considerable alarm when an H5N1 avian virus was found to have a very high mortality rate (~ 50%) in humans. However, sustained human-to-human transmission was not observed. The same happened in 2013 with an H7N9 virus. Clearly, other factors are necessary for effective spread in humans. During the first (Spring–Summer) phase of the 1918 pandemic, the mortality rate was low but it increased significantly during the second phase (Winter 1918/19). The cause of this increased pathogenicity is not known but is believed to be polygenic and would have

---

Box 28.1 THE GLOBAL INFLUENZA SURVEILLANCE AND RESPONSE SYSTEM (GISRS)

GISRS was established in 1952 to conduct global influenza surveillance. It is coordinated by the World Health Organization and endorsed by national governments. GISRS is a global network of 150 laboratories that monitor the types of influenza that are in circulation at any time in 114 countries around the world. More than 2 million respiratory specimens are tested by GISRS annually. The surveillance information that they collect is passed to the World Health Organization which collates it. They in turn advise vaccine manufacturers on the changes in influenza viruses circulating in the global population and what changes they need to make to maintain the efficacy of their seasonal influenza vaccines. GISRS has been considered to be effective in providing early warning of seasonal changes but occasionally the virus evolves in unexpected ways and the seasonal vaccines are less effective than usual.

---

arisen by antigenic drift. Experimental studies of influenza viruses in animals have indicated that the genes for HA, NA, the RNA polymerase, and the nucleoprotein contain potential virulence factors but these remain to be identified.

Influenza is not just a significant disease of humans, it also can cause major problems for the poultry industry and in this context it is worth noting that chicken is a major source of protein for much of the developed world. Outbreaks of highly pathogenic avian influenza occur from time to time in the poultry industry, sometimes involving as many as 1 million birds. Typically, the onset is rapid and a large percentage of the birds die within a few days. There are mandatory control measures in most Western countries which involve culling of entire flocks and bans on movement.

Today, vaccination is the primary defence against the morbidity and mortality of seasonal influenza (Box 28.1). Unfortunately, antibodies elicited by seasonal influenza vaccines do not provide protection when antigenic shift results in a virus with a novel HA component. Also, it has been difficult to rapidly scale up production of a new vaccine. Typically, it has taken six to nine months to generate enough stock for mass vaccination of the most vulnerable although this might change given the amazing response to vaccine production seen during the recent COVID-19 pandemic.

A number of antiviral drugs are available for treating influenza infections. Amantadine and rimantadine are antiviral drugs in a class of medications known as adamantanes that are best used prophylactically. They are active against influenza A viruses, but not influenza B viruses, but many seasonal strains that have circulated recently are resistant to them. A more effective class of drugs are the neuraminidase inhibitors such as the orally active oseltamivir (Tamiflu), inhaled zanamivir (Relenza), and the intravenously administered peramivir (Rapivab). They act by inhibiting the sialidase activity of the NA thereby inhibiting the release of progeny virions from virus-infected cells. Oseltamivir is the drug of choice for treatment of seriously ill individuals because it can be given orally but resistance can be a problem. Fortunately, there is not cross-resistance from oseltamivir to the other two neuraminidase inhibitors. Nevertheless, preventing influenza by vaccination rather than treating it, is the favoured control measure but the virus's segmented genome which permits reassortment will continue to challenge the vaccine manufacturers.

Key points

- Influenza is caused by a virus with a segmented RNA genome. The virus can infect humans, pigs and other mammals, and birds. The viral envelope contains two key proteins, haemagglutinin (HA) and neuraminidase (NA). There are sixteen known variants of the HA and nine of the NA.

- During an infection, HA binds to receptors on airway epithelial cells. Human influenza viruses and bird influenza viruses have different receptors. Some animals, such as pigs, have both types of receptors. NA removes sialic acid residues from viral glycoproteins and plays a role in the infection process.

- The RNA polymerase of influenza is error-prone and this results in mutations in HA and NA (antigenic drift). Some of these mutants are more infectious and cause seasonal influenza.

- If an avian influenza virus and a human influenza virus infect an animal with both virus receptors, such as a pig, reassortment of the eight RNA segments from each virus can occur. If any of the 256 possible new combinations (antigenic shift) has increased virulence then it can cause a pandemic.

- Seasonal vaccination of those most at risk is the best preventative strategy but occasionally the virus changes in unexpected ways and the vaccines in use have reduced effectiveness.

Suggested Reading

Lycett S.J., Duchatel F., and Digard P. (2019) A brief history of bird flu. *Philosophical Transactions of the Royal Society A* **374**, 20180257. doi:10.1098/rstb.2018.0257

Mostafa A., Abdelwhab E.M., Mettenleiter T.C., and Pleschka S. (2018) Zoonotic potential of influenza A viruses: a comprehensive overview. *Viruses* **10**(9), 497. doi:10.3390/v10090497

Short K.R., Kedzierska K., and van de Sandt C.E. (2018) Back to the future: lessons learned from the 1918 influenza pandemic. *Frontiers in Cellular and Infection Microbiology* **8**, 343. doi:10.3389/fcimb.2018.00343

29

Third Time Unlucky: SARS, MERS, and COVID-19

Box 29.1 CAVEAT

This chapter was written in April 2021 when the COVID-19 pandemic still was causing critical healthcare overload in many countries. Consequently, balanced retrospective analysis was not available and current ideas on the origin, spread, control and pathogenicity may turn out to be misplaced on later analysis. To make writing more difficult, 120,000(!) papers on COVID-19 had been published in the previous 12 months and picking out the most relevant was extremely difficult.

An outbreak of severe acute respiratory syndrome (SARS) began in November 2002 in Guangdong Province in China. One of those who was infected, a market vendor named Zhou Zuofen, checked into the Sun Yat-Sen Memorial Hospital in January 2003. He turned out to be a super-spreader (Box 29.2) and infected thirty nurses and doctors. Soon the disease had spread to neighbouring hospitals and by 10 February there had been 305 cases including 105 healthcare workers. Also in February, Liu Jianlun, a member of staff from the Sun Yat-Sen Memorial Hospital, travelled to Hong Kong to attend a family wedding and checked in to the Metropole Hotel. Despite feeling ill, he and his family travelled around Hong Kong visiting family. Soon he was hospitalized and subsequently died. He, too, appears to have been a super-spreader. Soon SARS was taken to Vietnam, Canada, Singapore, and Taiwan by people who were known to have stayed in the Metropole Hotel. By July 2003 there were no more detected infections but there had been 8,096 reported cases in 29 countries and 774 deaths, many in healthcare workers who nursed patients with SARS.

Due to a remarkable global effort, the cause of SARS was identified in April 2003. It was a coronavirus (CoV), one of a large group of viruses that

Microbiology of Infectious Disease. Sandy B. Primrose, Oxford University Press.
© Sandy B. Primrose (2022). DOI: 10.1093/oso/9780192863843.003.0029

Box 29.2 SUPER-SPREADERS

The *individual reproductive number* represents the number of secondary infections caused by a specific individual during the time that individual is infectious. Some individuals have significantly higher than average individual reproductive numbers and are known as super-spreaders. Through contact tracing, epidemiologists have identified super-spreaders in diseases as diverse as MRSA infections, tuberculosis, measles, rubella, monkeypox, smallpox, Ebola haemorrhagic fever, and SARS. What makes a particular person a super-spreader is not known but in the case of viral diseases there may be a relationship to the viral load in the infected person. In the case of people infected with respiratory diseases, such as COVID-19, a congested nose increases the velocity of respiratory droplets on coughing and sneezing. The thickness of a person's saliva also influences the spread of disease. Droplets formed from thick saliva rapidly fall to the ground but those from thin saliva stay airborne for long periods.

---

infect animals such as bats, birds, and mammals with occasional spillover to humans. Coronaviruses have a crown-like appearance under the electron microscope and are named after the Latin word 'corona', meaning crown or halo. They were first identified as human pathogens in the 1960s and, before SARS, four were known to cause mild to moderately severe human infections. Three of them (HCoV-OC43, HCoV-HKU1, and HCoV-229E) cause common colds and severe lower respiratory tract infections among people in the youngest and oldest age groups. The fourth (HCoV-NL63) is an important cause of (pseudo) croup and bronchiolitis in children. What made SARS so different and so worrying was its case fatality rate (9.5%). Fortunately, in the initial outbreak of SARS, hospitals had introduced strict quarantine procedures for both patients and staff treating them. This isolation of infection foci worked because the peak viraemia occurs five to ten days after the onset of symptoms. Another feature of the virus that helped manage its control is that it is not very transmissible except from super-spreaders.

In late 2003, SARS re-emerged in Guangdong Province. This time, the Chinese authorities acted fast by isolating suspect cases and in instigating contact tracing and quarantine measures. Civet cats were thought to be the source of this second SARS outbreak and so there was a mass cull of these cats in animal markets and breeding farms. This time, only five people were infected, and SARS was eliminated completely by January 2004 and has not been encountered since then. However, viruses very closely related to SARS have been found in bats and these viruses can infect human cells in culture without any adaptation in another host. This suggests that SARS could re-emerge.

In 2012, ten years after the emergence of SARS in China, a man in Saudi Arabia died of pneumonia and renal failure. A novel coronavirus was isolated from his sputum and the disease it caused was given the name Middle Eastern Respiratory Syndrome (MERS). A cluster of cases of severe respiratory disease had occurred in Jordan a few months previously and retrospectively were diagnosed as MERS. Then cases of MERS began appearing in other countries and all were linked through travel to, or residence in, countries in and near the Arabian Peninsula. More important, as with SARS, most of the cases were the result of nosocomial (hospital-acquired) infections. The largest known outbreak of MERS outside the Arabian Peninsula occurred in 2015 when an infected traveller returned to South Korea from the Arabian Peninsula; 16 hospitals and 186 patients were affected. Unlike SARS, MERS has not been eliminated and as of March 2021 there had been 2,580 reported cases from 27 countries with a case fatality rate of 33%. Fortunately, the MERS coronavirus does not efficiently transmit between humans and new cases usually arise from people being infected from its reservoir, dromedary camels which are present in high numbers in the Middle East.

Yet another coronavirus emerged in China in 2016. This originated in bats and caused a novel epizootic disease in pigs known as Swine Acute Diarrhoea Syndrome (SADS). The SADS virus is not closely related to SARS or MERS but is related to the NL63 and 229E coronaviruses that cause mild to moderate disease in humans. The occurrence of SADS led to an upsurge in Western exports of pigs to China and brought the disease to the attention of the agricultural community but not necessarily the public health community. Then, in late 2019, yet another coronavirus-associated disease arose in China: COVID-19. This was the third coronavirus to cause serious human disease since the beginning of the twenty-first century and, unlike its two predecessors, it was highly transmissible. As of late July, 2021 there had been 197 million cases and 4.2 million deaths. Although the average case fatality rate is only 2%, much lower than with SARS and MERS, there is a wide variation in the rate between countries with the highest rates occurring in lower income countries. For comparison, the 2009–2010 H1N1 influenza pandemic resulted in over 1.5 billion cases but only ~ 300,000 deaths (case fatality rate 0.02%). This explains why COVID-19 has caused global panic and the 2009–2010 influenza pandemic was barely noticed by the general public. The difference in case fatality rates between the different viruses most probably is correlated with the ability to cause cytokine storms (Box 29.3) in patients.

Box 29.3 CYTOKINE STORMS

There is no generally accepted definition of the term 'cytokine storm' but essentially it is an overstimulation of the innate immune system resulting in the uncontrolled and excessive release of the pro-inflammatory signalling molecules known as cytokines.

Patients who are severely ill with COVID-19 have high blood levels of the cytokines interleukin-6 and interleukin-1 and patients treated with antibodies to these cytokines had a lower mortality than those who were untreated. Interleukin-6 increases vascular permeability and causes fluid accumulation in the lung alveoli resulting in respiratory distress. There was high mortality in the 1918 influenza pandemic (p217) and the H1N1 virus reconstructed from victims of that pandemic has been shown to cause cytokine storms in mice.

Why is COVID-19 so highly transmissible when SARS and MERS were not? There may be multiple reasons but one factor that has been identified as being relevant is the structure of the viral spike, the S glycoprotein. This has two subunits, S1 and S2. The subunit S1 binds to its receptor on human epithelial cells which is the enzyme known as angiotensin-converting enzyme 2 (ACE2). SARS, but not MERS, also uses ACE2 as its receptor but its affinity for this receptor is ten to twenty times lower than that of COVID-19. Furthermore, the COVID-19 S glycoprotein is cleaved by host cell membrane proteases resulting in activation of the S2 subunit, a process that facilitates infection of cells by the virus. This cleavage does not occur with SARS.

One of the most remarkable aspects of the COVID-19 pandemic was the speed with which vaccines were developed and approved. Basically, three types of vaccine have been developed (Figure 29.1). The RNA vaccines from Pfizer–BioNTech and Moderna contain messenger RNA (mRNA) encoding the spike protein. The delivery of mRNA is achieved by a coformulation of the molecule into lipid nanoparticles which protect the RNA strands and help their absorption into the cells. The Oxford-Astra Zeneca and Janssen vaccines are non-replicating viral vector vaccines that use an adenovirus shell containing DNA that encodes the COVID-19 spike protein.

The emergence of four significant coronavirus epidemics (SARS, MERS, SADS, and COVID-19) in less than twenty years raises questions about their origins. What is clear is that they are zoonoses and, in the case of COVID-19, not the product of a biological warfare laboratory (Box 29.4). The viruses causing SARS, SADS, and COVID-19 have genome sequences that are very similar to those of coronaviruses found in Chinese horseshoe bats. Although the reservoir of the MERS virus is camels, this virus also is closely related to coronaviruses found in other species of bat. A major survey of more than 19,000 animals (bats, rodents, and non-human primates) in 20 countries spread over Asia, Africa, and the Americas found that 98% of the animals positive for coronaviruses were bats. More worrying, 9% of over 12,000 bats studied were carrying one or more coronaviruses. Given this, the occurrence of the COVID-19 pandemic is hardly surprising.

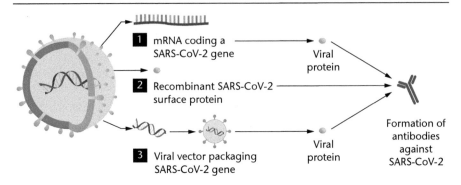

Figure 29.1. The three basic types of COVID-19 vaccine.

Source: Flickr/US Government Accountability Office (Public domain) <https://flickr.com/photos/58220939@N03/49948301838>.

The results of the coronavirus survey raise three important questions. Why have there not been more coronavirus epidemics in humans and domestic animals? Why have three of the recent coronavirus infections originated in China when infected bats are found worldwide? How can we prevent future coronavirus pandemics? We can only speculate about the answers to the first two questions. Disturbing the environment in which bats live will alter their behaviour and make them seek new habitats. This will bring them into contact with different animals. In the past few decades, China has undertaken massive construction projects and mass relocation of people. These activities could well have caused massive dislocation of bats. This, coupled with the widespread occurrence of wet markets where humans intermingle with large numbers of different animals, could well have promoted the transmission of coronaviruses to humans via intermediate hosts.

Preventing future outbreaks of coronavirus infections will not be easy. Population growth and industrialization means that bat habitats will continue to be destroyed. Shutting down wet markets would be a positive step and not impossible in an authoritarian country like China. Indeed, during the SARS re-emergence in 2003–2004 the Chinese did temporarily close local markets. Where such markets do continue to exist, and where there is much human contact with wild mammals, there should be a regular screening programme for coronaviruses. If coronaviruses are found then culling should be considered. This is what happens in Western countries when chicken flocks get infected with influenza (p221). Although prevention is better than a cure, the COVID-19 pandemic has taught us that there will be no universal coronavirus vaccine. Therefore, global efforts are needed to try and identify a suitable antiviral drug, preferably one that can be taken orally.

Box 29.4 DID COVID-19 ESCAPE FROM A LABORATORY?

Almost from the start of the COVID-19 pandemic there have been many theories about its origin, the most prevalent being that the virus is man-made and accidentally escaped from a Chinese laboratory. There has been a lack of openness on the part of the Chinese authorities and this has fuelled all sorts of conspiracy theories. What is known is that the Wuhan Institute of Virology had a team of virologists studying viruses isolated from bats. One of these viruses had caused serious illness in a group of miners who were harvesting guano from caves in Yunnan province. The miners became seriously ill with a form of pneumonia and three of them died. There are clear parallels here with the outbreaks of Marburg virus in Africa described on page XXX. The virus that killed the Chinese miners was a coronavirus and its nucleic acid sequence is similar to that of COVID-19.

The Wuhan laboratory also was undertaking research on what is known as 'gain of function' with the objective of increasing virus transmissibility and virulence. The aim of this work is to try to understand what makes a pandemic strain rather than to create a biological weapon. However, it is essential that such work is done under the highest biological containment conditions. Four biosafety levels are recognized internationally and they are designated BS1 to BS4. The work in Wuhan on novel bat viruses should have been done at level BS3 and the gain of function work at BS4 but it was not: it was done at biosafety level BS2. Based on the foregoing, the fact that the outbreak started in Wuhan where there is a laboratory that studies bat viruses strongly suggests that a laboratory-acquired infection and/or a virus leak of a new or engineered coronavirus that was under investigation was the origin of the COVID-19 pandemic.

Key points

- There have been three serious human diseases caused by coronaviruses since the beginning of the twenty-first century: SARS, MERS, and COVID-19. The case fatality rate with SARS and MERS was much higher than seen with COVID-19 but the infectivity was much lower. The increased mortality with SARS and MERS may be due to their propensity to cause cytokine storms in patients.

- COVID-19 and SARS use the same receptor (ACE2) on target cells but COVID-19 has ten times the affinity of SARS. This probably explains the much higher infectivity of COVID-19.

- A survey of large numbers of animals to identify potential reservoirs of coronaviruses found that 98% of infected animals were bats and that 9% of bats were carriers. Rapid industrialization in China with associated disturbance of bat habitats might explain the recent emergence of coronavirus infections. The existence of wet markets in China will exacerbate infection of humans.

Suggested Reading

Anderson R.M., Fraser C., Ghani A.C., Donnelly C.A., Riley S., et al. (2004) Epidemiology, transmission dynamics and control of SARS: the 2002–2003 epidemic. *Philosophical Transactions of the Royal Society B* **359**, 1091–105.

Anthony S.J., Johnson C.K., Greig D.J., Kramer S., Che X., et al. (2017) Global patterns in coronavirus diversity. *Virus Evolution* **3**(1), vex012. doi:10.1093/ve/vex012

De Wit E., van Doremalen N., Falzarano D., and Munster V.J. (2016) SARS and MERS: recent insights into emerging coronaviruses. *Nature Reviews Microbiology* **14**, 523–34.

Elrashdy F., Redwan E.M., and Uversky V.N. (2020) Why COVID-19 transmission is more efficient and aggressive than viral transmission in previous coronavirus epidemics. *Biomolecules* **10**(9), 1312. doi:10.3390/biom10091312

Gilbert S. and Greene C. (2021) *'Vaxxers*. Hodder & Stoughton.

Morens D., Breman J.G., Calisher C.H., Doherty P.C., Hahn B.H., et al. (2020) The origin of COVID-19 and why it matters. *American Journal of Tropical Medicine and Hygiene* **103**, 955–9.

Part V
Some Unifying Themes

30

Zoonotic Diseases

Zoonoses, infectious diseases in humans that arise from transmission from animals, are a consequence of humans living and working in close proximity to animals. A number of them have featured in earlier chapters: plague, anthrax, tuberculosis, salmonellosis, Lyme disease, malaria, Marburg virus, Ebola virus, influenza, SARS, MERS, and COVID-19. The significance of zoonoses can be seen from the statistics provided by the US Centers for Disease Control and Prevention (CDC): six in ten human cases of infectious disease arise from animal transmission and three out of four new or emerging infectious diseases. Indeed, the CDC is so concerned at the growing threat from zoonoses (Figure 30.1) that they have established the National Centre for Emerging and Zoonotic Infectious Diseases (NCEZID).

Zoonoses have been with us at least since humans adopted farming as a way of life but it is only in the last twenty-five years that they have become of significant concern to public health officials. In the 1970s it had been discovered that recombination between avian and human strains of influenza virus could give rise to pandemics (Chapter 28) but this was largely treated as being of academic interest. The change came with the detection of H5N1 influenza in patients in Hong Kong. Eighteen people were known to have been infected and six died—a very high rate of mortality. The Hong Kong government responded by culling the territory's entire poultry population, 1.5 million birds, within three days and this may have prevented a pandemic. The H5N1 strain resurfaced in humans, again in Hong Kong, in early 2003 when three people contracted the disease and two of them died. Between late 2003 and October 2005, 100 people in south-east Asia contracted H5N1 influenza and sixty died.

Three conditions must be met before an influenza pandemic begins: a new influenza strain that has not previously circulated in humans must emerge, this new strain must be capable of causing disease in humans, and the virus must be capable of being passed easily among humans. In the case of the

Microbiology of Infectious Disease. Sandy B. Primrose, Oxford University Press.
© Sandy B. Primrose (2022). DOI: 10.1093/oso/9780192863843.003.0030

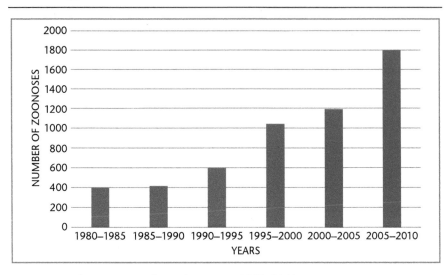

Figure 30.1. The rise in numbers of zoonoses 1980–2010.

H5N1 strain, transmission from human to human was not very efficient and only occurred as a result of very close contact with a patient during the acute phase of illness. If a more infectious variant was to arise then a pandemic would be triggered and, via the global influenza surveillance and response system (p221), public health functions around the world would be put on high alert. Another significant event occurred in 2003: an outbreak of severe acute respiratory syndrome (SARS) when 8,096 people were infected and 774 died. Like H5N1 influenza, SARS originated in birds (p225) and had a high case fatality rate (9.5%) but was poorly transmissible between humans. Even so, controlling it took an immense medical and public health effort. Again, had a more infectious variant arisen the consequences would have been a pandemic like that caused by COVID-19 but one with a far higher rate of mortality.

As the COVID-19 outbreak has shown us, once a pandemic starts it is extremely difficult to bring it under control. Prevention is a much better strategy but this means having adequate surveillance systems in place and having suitable plans for containment. The H5N1 influenza and SARS outbreaks were the wake-up call for the global public health community as to the importance of controlling zoonoses. There were two outcomes: the establishment of the International Health Regulations (Box 30.1) and the establishment of One Health. The latter is a collaborative and transdisciplinary approach that works at local, regional, national and global levels. One Health recognizes that the health of people is closely connected to the health of animals *and* our shared environment. The importance of the

---

Box 30.1 INTERNATIONAL HEALTH REGULATIONS

The International Health Regulations (IHR) are an international legal instrument that is binding on 196 countries across the globe that came into force on 15 June 2007. The aim of the regulations is to help the international community prevent and respond to acute public health risks that have the potential to cross borders and threaten people worldwide. As part of IHR implementation, World Health Organization (WHO) member states are committed to strengthening their ability to rapidly detect, assess, and report potential public health emergencies of international concern (PHEICs).

Events are identified as potential PHEICs and reported to the WHO when they fulfil at least two of the following four criteria:

- a serious public health event is suspected;
- the event is considered unusual or unexpected;
- there is a significant risk of international spread;
- the event poses a significant risk to international travel or trade.

Events that have been reported or detected as potential PHEICs are entered into a database known as the Event Management System, which is administered by WHO.

The effectiveness of the IHR depends on the speed with which national authorities notify WHO of a PHEIC. COVID-19 meets all four of the above criteria for a PHEIC but China delayed notifying the international community and the result was the 2020–2021 pandemic.

---

environment cannot be overemphasized. The rampant deforestation, uncontrolled expansion of agriculture, intensive farming, mining, and infrastructure development of the last fifty years has had three consequences, particularly in poorer countries. First, wildlife now is encroaching on areas inhabited by humans creating opportunities for spillover of diseases from wildlife to people. Second, the reduction in biodiversity and allows smaller animals such as bats and rats, who are more adaptable to human pressures and also carry the most zoonotic diseases, to proliferate. Third, deforestation in tropical and sub-tropical areas leads to a huge expansion in the numbers of mosquitoes and other biting insects and hence the transmission of disease. A good example was the huge rise in cases of malaria during the Vietnam War (p168) resulting from aerial bombardment with herbicides.

Disease transmission from animals can occur even without environmental destruction because animals play an important role in our lives, including for food, clothing, sport, and companionship. As an example, during the lockdowns associated with the COVID-19 pandemic, many more people than usual in the United Kingdom acquired dogs as pets. Many of these dogs were sourced from unofficial suppliers and a significant number have brucellosis, a disease that can spread to humans and cause serious illness. Another example is the international trade in animals and animal products that can result in diseases spreading far beyond their natural habitat. We have already discussed the outbreak of Marburg virus in Germany (p202) following the importation of African Green monkeys for research purposes. Many people in west Africa have a love of bushmeat and this led to the emergence of HIV/AIDS. More worrying is that when people from west Africa move to European countries, they try to satisfy their desire for bushmeat by importing it illegally, a practice that led to a UK outbreak of foot and mouth disease.

The One Health initiative and the IHR represent the two sides of a coin. The One Health objective is to try to eliminate existing zoonoses and prevent new ones from arising whilst the objectives of the IHR are to get early warning of potential zoonotic outbreaks and take appropriate control measures. Implementation of the IHR requires effective surveillance coupled with accurate and reliable reporting mechanisms. Surveillance in turn requires the availability of appropriate diagnostics, that is:

- tests that are cheap enough to be used on a mass scale, particularly in developing countries;
- tests that are simple to use and robust;
- tests that measure the right analytes.

Meeting these criteria for diagnostic tests is not easy and the recent COVID-19 pandemic illustrates the problem. Diagnosing potentially infected individuals was done using lateral flow tests, ELISAs (Enzyme Linked Immunoabsorbent Assays), and by quantitative PCR (polymerase chain reaction), and the analytes being determined varied from viral RNA, viral antigen, and antibodies to the virus. Lateral flow tests can be made very cheaply, are robust, and almost anyone can be trained to use them properly in a short space of time. Quantitative PCR is at the other end of the scale: it requires instrumentation, skilled staff, and relatively expensive consumables. There is another confounding factor. The sensitivity and specificity of diagnostic tests is crucially important as one wants the minimum number of false negative and false positive results. However, the sensitivity and specificity are very dependent on the viral load in infected individuals, the reagents used, the analyte being measured and the ability to detect all genotypic (PCR tests) or antigenic variants (lateral flow and ELISA tests).

The greatest biodiversity occurs in south-east Asia, sub-Saharan Africa, and the Amazon basin and they are the source of almost all of the zoonoses of concern in the last fifty years. These are the countries where surveillance needs to be greatest but all too often, they are the poorest countries. So, what screening should they do, given that the World Health Organization has a list of over 200 zoonoses? Another problem is that new zoonotic diseases are being discovered continually but may not appear as a serious public health impact at the time. Public health officials might not even know about new zoonotic diseases if they first appear in remote communities. For example, the Zika virus and the Ebola virus both appeared in small villages in Africa, but it was some decades before the diseases appeared in large cities and became public health emergencies of international concern.

The One Health initiative aims to understand how factors such as changes in land use and climate change can lead to changes in animal behaviour that will result in disease and what needs to be done to prevent the emergence of zoonoses both old and new. Should surveillance detect an increase in a zoonoses then appropriate control measures need to be put in place. In an ideal situation, infected individuals are identified as early as possible, quarantined, and cured of the disease, but achieving this is extremely difficult. Humans infected with bacterial, fungal, and some protozoan zoonoses can be treated with antibiotics but the cost of these can be prohibitive to many developing countries. However, most of the recent zoonoses of concern have been viral and there are very few effective antiviral drugs. Treatment of infected domestic animals is usually not practical and so culling is the preferred option even though there is significant economic cost. However, if the reservoir of a zoonotic infection is a wild animal there really are no effective options. Prevention of disease by use of vaccines is preferable to treatment but, cost apart, effectiveness depends on the type of infectious agent being targeted. It is difficult to get long-lasting prophylaxis with bacterial vaccines but this is much less of a problem with viruses for which the therapeutic options are limited.

COVID-19 has been the most devastating PHEIC in terms of human and economic cost since the 1918–1919 influenza pandemic. Despite the existence of the IHR and One Health, the world managed it very badly. It remains to be seen what changes will be implemented both nationally and internationally.

Key points

- The number of zoonoses of global concern has increased every decade for the last sixty years and is linked to habitat destruction and biodiversity loss.

- Most countries in the world have signed up to the International Health Regulations (IHR) whose objective is to prevent and respond to potential public health emergencies of international concern (PHEIC). A key issue with the IHR, as exemplified by COVID-19 is the willingness of a country to report a PHEIC.

- Good surveillance is essential for the early detection of zoonoses that may become a PHEIC but the availability and cost of appropriate diagnostics is a limiting factor, particularly in the poorer countries which often have the greatest reservoirs of zoonoses.

- The One Health initiative, which operates at all levels from local to global, has the objectives of trying to prevent the emergence of new zoonoses and responding effectively to any PHEIC that occurs.

Suggested Reading

Cross A.R., Baldwin V.M., Roy S., Essex-Lopresti A.E., Prior J.L., and Harmer N.J. (2019) Zoonoses under our noses. *Microbes and Infection* **21**, 10–19.

Gibb R., Redding D.W., Chin K.Q., Donnelly C.A., Blackburn T.M., et al. (2020) Zoonotic host diversity increases in human-dominated ecosystems. *Nature* **584**, 398–402.

McAleenan B. and Nicolle W. (2020) Outbreaks and Spillovers. doi: https://policyexchange.org.uk/wp-content/uploads/Outbreaks-and-Spillovers.pdf

Recht J., Schuenemann V.J., and Sánchez-Villagra M.R. (2020) Host diversity and origin of zoonoses: the ancient and the new. *Animals* **10**(9), 1672. doi:10.3390/ani10091672

Schneider M.C., Munoz-Zanzi C., Min, K-D., and Aldighieri S. (2019) 'One Health' From Concept to Application in the Global World doi.org/10.1093/acrefore/9780190632366.013.29

Waltner-Toews D. (2017) Zoonoses, One Health and complexity: wicked problems and constructive conflict. *Philosophical Transactions of the Royal Society B* **372**, 2016171. doi: 10.1098/rstb.2016.017

31

Some Common Pathogenicity Themes

When I was an undergraduate in microbiology nearly sixty years ago, much teaching time was devoted to a study of pathogens. However, many of the pathogens of concern today had not even been discovered then, for example, *Legionella pneumophila*, *Helicobacter pylori*, *Borrelia burgdorferi*, as well as *Campylobacter* species (not discussed herein), or the diseases caused by HIV, Marburg virus, Ebola virus, SARs. MERS, and COVID-19 and a host of other viruses. Three factors have contributed to the rise of new diseases. First is the emergence of zoonoses through habitat destruction as detailed in Chapter 30. Interestingly, even though the term 'zoonosis' was coined at the end of the nineteenth century, I do not recall ever hearing the word in my years as a student. Second is the exponential growth in air travel since the late 1950s coupled with a trend for Westerners to take holidays in 'exotic' locations. Third is the huge increase in the numbers of immunocompromised individuals. The immunosuppressive drug cyclosporin was discovered in 1971 and entered clinical use in 1983 for the prevention of graft-versus-host rejection in bone marrow transplants and rejection of organ transplants. The number of individuals who are immunocompromised as a result of taking cyclosporin can be judged from a single statistic: in 2017, over 1 million prescriptions for it were filled in the United States alone. Another major factor influencing the number of immunocompromised individuals is the exponential increase in the use of chemotherapy to treat various cancers, a factor exacerbated by the nearly thirty-years increase in lifespan in developed countries over the last sixty years. Last, but not least, is the number of individuals who are immunocompromised as a result of HIV/AIDS: 38 million at the last count of whom nearly 2 million were children.

Again, at the time of my university education, none of the key themes relating to pathogenicity that are discussed in this book were known, for example, horizontal gene transfer, secretion systems and associated effectors, PAMPs, and pathoadaptation. Cellular immunology was in its infancy

Microbiology of Infectious Disease. Sandy B. Primrose, Oxford University Press.
© Sandy B. Primrose (2022). DOI: 10.1093/oso/9780192863843.003.0031

and DNA sequencing was twenty years away. Nor had molecular clocks been conceived so there was nothing to suggest that some diseases such as canine distemper, measles, cholera, and Neisserial meningitis were only a few hundred years old.

For the viruses discussed in this book there are two key themes. First, most of the viruses with pandemic potential have an RNA genome, rather than a DNA one, and replicate with the aid of an RNA-dependent RNA polymerase. As this polymerase has no proof-reading function it is error-prone and this provides a source of mutations that can increase the fitness of the virus. An additional source of variation is provided if, like influenza, the virus has a segmented genome. The animal kingdom is replete with viruses that often cause few if any symptoms in their usual host, particularly bats (p205 and p228). However, when such viruses cross the species barrier into humans then serious disease can occur. The second key theme is that, fortunately, there are many barriers to successful spillover. However, error-prone replication can generate mutants that can overcome these barriers. The habitat destruction that is going on in many parts of the world is going to increase the likelihood of spillover, and not just of coronaviruses (p227).

Horizontal gene transfer has played a role in the evolution of almost all the cellular pathogens described in this book. Whilst the role of plasmids has been highlighted it is worth noting that lysogenic conversion occurs frequently as well. Notable examples of phage-borne toxin genes can be found in *Vibrio cholerae*, *Pseudomonas aeruginosa*, *Escherichia coli*, *Staphylococcus aureus*, *Streptococcus pyogenes*, *Clostridium botulinum*, *Corynebacterium diphtheriae*, and the plant pathogen *Ralstonia solanacearum*. Even when they do not carry toxin genes, bacteriophages can increase the pathogenicity of bacteria as has been observed in the meningococcus (p71) and *Pseudomonas aeruginosa* biofilms in cystic fibrosis (p84).

Plants and animals have evolved Pattern Recognition Receptors that allow them to recognize PAMPs, the molecular motifs on microbes that are recognized by plants and animals and initiate defensive responses. In turn, the pathogens have acquired a variety of secretion systems that allow them to insert effectors into plants and animals to overcome the defensive strategies invoked by their binding to Pattern Recognition Receptors. Another, somewhat surprising, general mechanism of pathogenicity is the increase in virulence that occurs by mutations in genes that confer a growth advantage outside of the body, that is, pathoadaptation. The best example occurs in *Shigella* species (p35) but there are many others.

Finally, the role of insects in infectious disease should not be underestimated. Many plant viruses are transmitted mechanically from an infected plant to an uninfected plant. That is, the mouthparts of the insect get contaminated when feeding on an infected plant and some of this contamination gets transmitted to new hosts. However, in the case of some infectious diseases of humans, such as plague, Lyme disease, phytoplasmas, pathogenic protozoa like malaria, and some viruses, the pathogen undergoes a developmental stage in its insect host that is key to successful infection of a new host.

Only a very few microbial pathogens have been covered in this book and they were selected because they exhibit one or more of the features described earlier. Pick any other pathogen and they almost certainly will display some of the same features.

Suggested Reading

Brüssow H., Canchaya C., and Hardt W-D. (2004) Phages and the evolution of bacterial pathogens: from genomic rearrangements to lysogenic conversion. *Microbiology and Molecular Biology Reviews* **68**, 560–602.

Geoghegan J.L. and Holmes E.C. (2018) Evolutionary virology at 40. *Genetics* **210**, 1151–62.

Index